Elementary Statistics Laboratory Manual
Macintosh® Version

John D. Spurrier, Don Edwards, and Lori A. Thombs
University of South Carolina

An Alexander Kugushev Book

Duxbury Press
An Imprint of Wadsworth Publishing Company

I(T)P™ An International Thomson Publishing Company

Belmont • Albany • Bonn • Boston • Cincinnati • Detroit • London • Madrid • Melbourne
Mexico City • New York • Paris • San Francisco • Singapore • Tokyo • Toronto • Washington

Editorial Assistant: Janis Brown
Production Editor: Karen Garrison
Print Buyer: Barbara Britton

Cover: Stuart Paterson/Image House
Printer: Malloy Lithography

COPYRIGHT © 1995 by Wadsworth Publishing Company
A Division of International Thomson Publishing Inc.
I(T)P The ITP logo is a trademark under license.

Printed in the United States of America
3 4 5 6 7 8 9 10—01 00 99 98 97

The Apple logo , ImageWriter, and Macintosh are registered trademarks of Apple Computer, Inc. Minitab is a registered trademark of Minitab, Inc.

For more information, contact Wadsworth Publishing Company:

Wadsworth Publishing Company
10 Davis Drive
Belmont, California 94002, USA

International Thomson Editores
Campos Eliseos 385, Piso 7
Col. Polanco
11560 México D.F. México

International Thomson Publishing Europe
Berkshire House 168-173
High Holborn
London, WC1V 7AA, England

International Thomson Publishing GmbH
Königswinterer Strasse 418
53227 Bonn, Germany

Thomas Nelson Australia
102 Dodds Street
South Melbourne 3205
Victoria, Australia

International Thomson Publishing Asia
221 Henderson Road
#05-10 Henderson Building
Singapore 0315

Nelson Canada
1120 Birchmount Road
Scarborough, Ontario
Canada M1K 5G4

International Thomson Publishing Japan
Hirakawacho Kyowa Building, 3F
2-2-1 Hirakawacho
Chiyoda-ku, Tokyo 102, Japan

All rights reserved. No part of this work covered by the copyright hereon may be reproduced or used in any form or by any means—graphic, electronic, or mechanical, including photocopying, recording, taping, or information storage and retrieval systems—without the written permission of the publisher.

Library of Congress Cataloging-in-Publication Data
Spurrier, John D.
    Elementary statistics laboratory manual, Macintosh version / John D. Spurrier,
    Don Edwards, and Lori A. Thombs
    p.   cm.
    Includes index.
    ISBN: 0-534-23610-3
    1. Statistics—Data processing—Laboratory manuals. I. Edwards, Don. II. Thombs,
    Lori A. III. Title.
QA276.4.S63   1995
519.5'078–dc20   94-30084

DEDICATION

To Pam, Katie, and Ryan for their love, enthusiasm, and understanding
To Kris, the light at the end of the tunnel
To Norma and Howard

What I hear, I forget.
What I see, I remember.
What I do, I understand.
Chinese proverb

CONTENTS

Preface **vii**

Working in Teams: Welcome to the Real World **1**

Session 1 An Introduction to Macintosh and to Minitab **3**
Statistical Concepts: Dotplots, stem-and-leaf plots, outliers

Session 2 Variability and Heart Rates **29**
Statistical Concepts: Variability within and among individuals, effect of sample size, preliminary study, histogram, boxplot, numerical descriptive statistics

Session 3 Author, Author **47**
Statistical Concepts: Classification, by-treatment dotplots, scatter plots

Session 4 Secrets Behind a Green Thumb **67**
Statistical Concepts: Planned experiments, two-factor design, factor-level selection, randomization

Session 5 Real and Perceived Distances **79**
Statistical Concepts: Scatter plot, regression, calibration, bias, measurement error, variability within and among individuals

Session 6 Collecting Data over Time **95**
Statistical Concepts: Time series, trends, cycles, seasonal variation, smoothing a time series

Session 7 A Question of Taste **117**
Statistical Concepts: Binomial distribution, proportion, single-blind experiment, randomization

Session 8 Sampling and Variation in Manufactured Products **129**
Statistical Concepts: Variation, sources of variation, outliers, descriptive statistics, sampling distribution

Session 9 Exploring Statistical Theory Through Computer Simulation **147**
Statistical Concepts: Sampling distributions, comparison of estimators, sample mean, sample median, central limit theorem, simulation

Session 10 Improving Product Performance with Planned Experiments **167**
Statistical Concepts: Planned experiment, two-factor design, factor selection, randomization, interaction

Session 11 Breaking Strength of Facial Tissue   **185**
Statistical Concepts: Stem-and-leaf plot, normality assumption, one sample t-test and confidence interval, nonparametric sign confidence interval

Session 12 Conclusion of Plant-Growth Experiment   **201**
Statistical Concepts: Two-factor design, treatment combinations, multiple scatter plots

Session 13 The Race to Solution   **217**
Statistical Concepts: Two-factor design, factor selection, treatments, interaction, descriptive statistics

Session 14 Walk This Way   **235**
Statistical Concepts: Blocking, paired-sample *t* confidence intervals and hypothesis tests

Session 15 Absorbency of Paper Towels—A Messy Data Problem   **255**
Statistical Concepts: Population, census, simple random sample, systematic random sample, side-by-side boxplots, independent samples t-test and confidence interval

Session 16 Variation in Nature   **271**
Statistical Concepts: Scatter plots, simple linear regression, correlation

Session 17 Random Sampling   **293**
Statistical Concepts: Random sample, population, sampling frame, proportion, variability of a sample statistic, confidence interval for a proportion

Appendix 1 Technical Report Writing   **305**

Appendix 2 Technical Report Writing Checklist   **317**

Index   **321**

PREFACE

OUR OBSERVATIONS

Students in beginning statistics courses often view experimentation and data merely as words and numbers in a text. They plug numbers into formulas and then make conclusions about briefly described experiments. At times, students are asked to do some of these routine computations on a computer.

We observe that students generally complete these courses with the false impression that the field of statistics and a career in statistics deal solely with evaluating formulas and that statistical reasoning does not enter an investigation until after the data has been collected. Students generally receive little or no exposure to the important statistical activities of sample selection, data collection, experimental design, searches for sources of variation, the development of statistical models, the selection of factors, and so on. In short, they often leave the first course without understanding the role of statistics and the statistician in scientific investigations. They often leave the course thinking that statistics is dull.

OUR GOAL

Our primary goal in writing this laboratory manual is to lead students through a series of "hands-on" experiments that illustrate important points of applied statistics. We are attempting to give students examples and experiences that they will remember long after the course is over. We find that students participating in such experiments gain a more accurate impression of the role of statistics and the statistician in scientific investigations and a greater appreciation for science. They also tend to think about statistics at a deeper level than do traditional elementary statistics students. For example, difficult concepts such as outliers, interaction, the need for randomization, and the choice of predictor variables arise naturally and are understood as the students work through the experiments. Students also see that statistics is a fun and exciting field of study.

Our secondary goals are to improve students' technical writing and teamwork skills. These skills, which are highly valued by most employers, are not traditionally emphasized in elementary statistics courses.

DESCRIPTION OF LABORATORY SESSIONS

The experiments are designed such that the data can be collected with reasonably inexpensive measuring equipment. The data is analyzed using Minitab on a Macintosh computer. The sessions were developed using Minitab Release 8.2. All sessions except for Session 9 can be done using the Student Edition of Minitab. No previous experience with Minitab or a Macintosh is required. The level of mathematical maturity required is consistent with that expected for a traditional, noncalculus-based, elementary statistics course. Session 1 gives an introduction to the Macintosh and to Minitab. Sessions 2 through 17 guide the students through a series of experiments. The experiments do not depend on each other except for Sessions 4 and 12. Complete instructions are given for the use of Minitab in each session.

Each session includes a short answer writing assignment and an extended writing assignment. The short answer writing assignment can be completed and graded reasonably quickly. The extended writing assignments, which are more time-consuming, are designed to produce formal reports. The appendixes are particularly helpful for students in the preparation of extended writing assignments. We have found that using extended writing assignments for three sessions and short answer writing assignments for the rest of the sessions is a good mix. Additional instructions for running the session are given in the Instructor's Resource Manual.

PLACEMENT OF THE LAB IN THE CURRICULUM

We have run the laboratory as an optional fourth credit hour of our traditional three-semester-hour elementary statistics course. The lab meets weekly for two hours, similar to a biology or chemistry lab. The laboratory also could be used in lieu of some lectures in a three-semester-hour course. Alternatively, the laboratory could be used with upper-division courses in applied statistics, engineering statistics, or mathematical statistics. Some of the experiments have been used as course projects in settings where operation of the entire laboratory was not practical.

ACKNOWLEDGMENTS

The authors acknowledge the support of the National Science Foundation's Undergraduate Course and Curriculum Development Program through grant number USE-9155850 and the University of South Carolina in funding the project that led to this laboratory manual. Any opinions, findings, and conclusions or recommendations expressed in these materials are those of the authors and do not necessarily reflect the views of the National Science Foundation. We thank Sneh Gulati, Scott Goode, John Grego, Rhonda Grego, Robert Hogg, Ralph Johnson, Judith Jones, Leslie Jones, James Lynch, Donald Miller, Rosemary Oakes, Mary Ellen O'Leary, W. J. Padgett, Richard Scheaffer, Jennifer Simsick, Pamela Spurrier, Julie Starks, and numerous students for helpful suggestions that have greatly improved the experiments and the manual. We appreciate Alex Kugushev of Duxbury Press for his enthusiastic support of our work and the other Duxbury staff who have made the publication of our manual a reality. We thank Laura Gooding, Alphonso Mason, La Metis Johnson, Katie Spurrier, and Ryan Spurrier for performing early trial runs of many of the experiments. Laura Gooding, La Metis Johnson, and Jennifer Simsick also helped with the preparation of the graphics. We also thank Apple Computer, Inc., and Minitab, Inc., for granting permission to reproduce certain graphic images in this manual. Finally, we thank the following reviewers for numerous helpful comments during the development of the manuscript: Kenneth M. Brown, Jr., College of San Mateo; K. Ruben Gabriel, University of Rochester; H. Joseph Newton, Texas A & M University; Adolph A. Oliver, III, Chabot College; Sister Adele Marie Rothan, CSJ, The College of St. Catherine; and Jessica Utts, University of California, Davis.

The authors would appreciate any feedback from users of this manual. Written comments can be sent to any of the authors at the Department of Statistics, University of South Carolina, Columbia, SC 29208.

John D. Spurrier, Don Edwards, Lori A. Thombs

Working in Teams: Welcome to the Real World

Most of your work in collecting and analyzing data in the laboratory will be done as a member of a student team. Project teams are common in the workplace, and the ability to work in teams is a skill that most employers highly value. Use your experiences in this lab to help develop these important skills!

You will find that there are differences among your team members. They may differ in terms of academic major, work experience, computing experience, gender, race, and so on. These differences tend to give various team members different insights into the problems being discussed. Effectively combining these insights usually produces better solutions.

The following suggestions will help your team:

1. Each team member should be an equal partner. Do not fall in the trap of either trying to make all of your team's decisions or allowing other team members to do all of the work.

2. Respect the opinions of each team member. Other team members may have an insight into the problem that you do not have. You do not have to agree with them, but respect their right to express an opinion different from yours.

3. Respect the differences among team members. Nothing destroys teamwork quicker than a thoughtless remark belittling a person because he or she is different from you.

4. Communication involves listening and talking. Do not allow one person to dominate the conversation. Encourage everyone to share their thoughts.

5. Make decisions by reaching a consensus. When team members disagree, work out a solution that everyone can support. Recognize that this solution may not be anyone's first choice.

6. Make sure all team members understand what will be measured, how it will be measured, and who will record the data. Confusion regarding data collection leads to mistakes, misleading conclusions, and frustration.

SESSION ONE

An Introduction to the Macintosh® and to Minitab®

INTRODUCTION

You will be using Macintosh computers with a special statistics application program called Minitab for the data entry, graphics, and analyses. This session gets you started with the Mac and with Minitab.

STATISTICAL CONCEPTS

Dotplots, stem-and-leaf plots, outliers.

MATERIALS NEEDED

A blank diskette for each student.

THE SETTING

You are a teacher for a college honors course; the term is half finished. Your class has taken two exams and you would like to look closely at the mean test score distribution to assess their performance. You have to let the students know where they stand.

BACKGROUND

An **outlier**, loosely defined, is a very unusual value in a data set. Outliers are sometimes caused by mistakes, such as incorrectly reading an instrument or making an error when typing data into the computer. However, outliers are not always mistakes. Understanding the cause of an outlier can lead to important discoveries.

HOW TO USE A MACINTOSH

Follow the tutorial instructions below very carefully. First, move the **mouse** (the small box with a button on the pad next to the Mac) a little to see whether the Mac is already on—if the screen stays blank, the Mac is off. If you are using a small-screen Mac like the one shown in Figure 1.1, an on-off switch is at the left on the back. Big-screen Macintoshes usually have an unlabeled "on" key at the upper right on the keyboard.

Figure 1.1 Small-Screen Macintosh

After the Mac is on, you should see the **main menu**, which consists of ®, **File**, **Edit**, **View**, and **Special**, across the top of the screen. Figure 1.2 shows the main menu. This main screen is sometimes called the **desktop**. At the bottom right of the screen is a picture of a trash can. Little pictures like the trash can are called **icons**. At the top right may be a rectangle or something like it with a name below it. This is the icon and name for the computer's **hard disk drive**, which is where the computer's long-term memory is. This manual assumes that the Mac you're using has a hard disk drive. If not, your instructor will provide modifications to the instructions as needed.

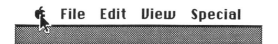

Figure 1.2 Main Menu

Your main control for working with a Macintosh is the mouse. Just as you had to practice with the controls of a car before you took it out on the freeway, you will need to develop some basic mouse skills to work effectively on the Mac. These skills are **pointing**, **clicking**, **dragging**, and **selecting menu items**. If you are in groups on the computer, each member of the group should learn pointing, then each member should learn clicking, and so forth. It will only take a few minutes; don't get in a hurry.

POINTING

Hold the mouse lightly between your thumb and middle finger. Move the mouse around a bit on the pad to see how it affects the arrow, or pointer, on the screen. If you get to the edge of the pad but want to move the pointer farther in that direction, just pick up the mouse and place it anywhere else on the pad and keep moving from there. Now, without pushing the mouse button, practice pointing by moving the pointer until it just touches each of the objects on the screen. Point to each corner of the screen in turn, in clockwise order and then in counterclockwise order.

CLICKING

Clicking involves putting the pointer on an icon or a menu item and pushing and then releasing the mouse button. Try clicking on the . Did you see a list appear and then disappear? That was the **menu**. If you want to look at the menu longer, point at the again and, this time push and hold down the mouse button. When you let the mouse button go, the menu disappears.

Release the mouse button now and click on the trash can. The trash can should now be highlighted. If you release the mouse button, the highlighted item is **selected**. Now click twice, as quickly as you can, on the trash can. This is called **double-clicking**. The empty box that opens on the screen is called a **window**, specifically the **Trash window**. Figure 1.3 illustrates the Trash window. We'll work with windows more later, but for now close the Trash window by clicking on the little box in the window's upper-left corner, which is called the **close box**.

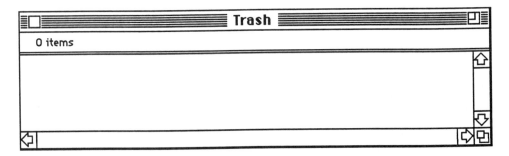

Figure 1.3 Trash Window

In summary, there are at least three variations on clicking:

Click Point to an object and push and then release the mouse button.

Click and hold Point to an object and push and then hold down the mouse button.

Double-click   Point to an object and push the mouse button twice, quickly.

DRAGGING

Click and hold on the trash can and, while still holding down the mouse button, move the mouse on the mouse pad. An outline of the trash can should move as though the trash can is pasted to the pointer. Drag the trash can over to the lower-left corner of the screen. Now drag it back. Drag it clockwise to all four corners of the screen. Now drag it counterclockwise. Finally, drag the trash can back to the lower-right corner of the screen where it belongs.

SELECTING MENU ITEMS

Click and hold on the . You'll see the   **menu**, which is a list of mini-programs called **desk accessories** that you can use for simple things such as checking the time or doing a calculation. Let's check the time. With the mouse button held down, move the pointer slowly down the screen from the . Notice that each menu item is highlighted when the item is touched by the pointer. Highlight the **Alarm Clock** and then, without moving the mouse, release the mouse button. You have just **selected** the Alarm Clock, and you should now have a small box similar to Figure 1.4 on the desktop showing you the time. Close this desk accessory now by clicking in its close box (at the left of the time shown).

Figure 1.4 Alarm Clock

Now select the **Calculator**, another desk accessory, from the   menu. That is, highlight the Calculator menu item and then carefully release the mouse button. You should see a picture of a calculator similar to Figure 1.5 on the screen. You can push buttons on this calculator by clicking on them. Try it; click on each of these in turn: 4, *, 8, =. The calculator register should show you 32, the answer to 4 times 8. The "C" button clears the calculator—try clicking on it now. You can also enter numbers on the calculator from the numeric keypad on the right side of the keyboard. Try using the numeric keypad to compute 217 divided by 67.3. It's 3.224, right? Now click on the

Calculator's close box at the upper-left corner to put the Calculator away. The Alarm Clock and Calculator are two desk accessories you may want to use from time to time.

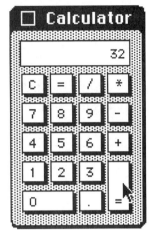

Figure 1.5 Calculator

Congratulations! If you accomplished all the above, you're now certified as a safe (well, maybe not safe) beginning Macintosh user. A couple more beginning topics are needed before we introduce the data analysis program, Minitab.

MORE ABOUT WINDOWS

Nearly all Macintosh work is done in windows of one sort or another. These windows have several characteristics in common. As an example, open the Trash window again by double-clicking on its icon. Across the top of every window is the **title bar**. You can drag the window anywhere you want on the desktop by clicking and holding on the title bar and then dragging the window. Try this by dragging the open Trash window to the top of the screen.

There are two ways to resize a window:

1. Click in the box at the top right of the window and the window gets very big. Click again and the window returns to its original size.

2. Click and hold in the box at the lower right of the window (the **resize box**) and then drag. The window changes shape and size depending on where you drag that corner.

The pointer in Figure 1.6 indicates the resize box. We'll talk about the bars on the right and bottom of the window later. Close the Trash window now by clicking in its close box.

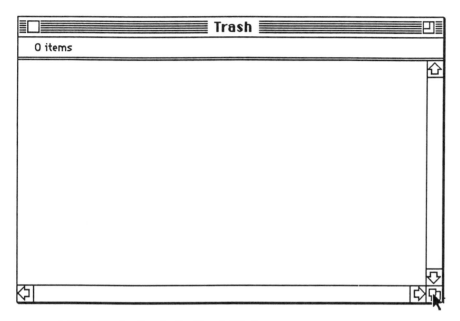

Figure 1.6 The Resize Box in the Trash Window

DISKETTES

Diskettes are the plastic disks, 3-1/2 inches square, that go in the slot in the front of your Mac (this slot is the **internal disk drive**). Note that there are several different kinds of diskettes with varying storage capacities. Your diskette is probably a **high-density** diskette. If so, it has an HD logo next to the metal part of the diskette, as shown in Figure 1.7. It has very high storage capacity, but old Macintoshes such as 512 Kb Macs, Mac Pluses, and some Mac SEs and Mac IIs cannot read HD diskettes. They require another kind of diskette, called a **double-density** diskette. A double-density diskette will not hold as much information as an HD diskette, but both old and new Macintoshes can use it. If you will be using older Macs in your work, you will want to purchase a double-density diskette at a bookstore or office supply store.

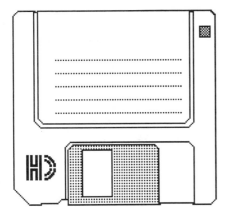

Figure 1.7 High-Density Diskette

FORMATTING A DISKETTE

Insert your diskette, metal side first and label side up, into the internal disk drive on the front of the Macintosh. Whenever you insert a new diskette into a drive, you'll have to **format**, or **initialize**, the diskette as follows:

1. A **dialog box** will open on the screen. The dialog box says, "This disk is unreadable. Do you wish to initialize it?" Click on the button labeled **two-sided**.

2. Another box will open, warning you that initializing will erase all information on the diskette. Since there is no information on a new disk anyway, click **erase**.

3. Another box will open that says "Please name this disk." On the keyboard, type a name; we recommend you use your own name along with the course number or name. If you make a mistake typing, push the Delete key at upper right on the keyboard until you erase the mistake. *Throughout this manual, it will be convenient to refer to keys on the keyboard by putting the name of the key in braces.* Thus, <Delete> will denote the Delete key.

4. When the name of the diskette is correct, click the **OK** button.

After some delay, the initializing process will end and you will be able to use your diskette from then on.

    Whenever a diskette is in the internal disk drive, that diskette's icon and name are shown at upper right on the desktop, below the hard disk drive's icon. When you want to remove the diskette you may **eject** it by dragging its icon to the trash can. Practice ejecting the diskette now.

If you are sharing a computer, initialize each team member's diskette now. Then insert one member's diskette and keep it in the disk drive throughout each session's computer work. That way, when it comes time to save data sets, plots, and other information a diskette will be in the disk drive, ready and waiting. After each session, you can copy information from your partner's diskette to your own, or vice versa, as described at the end of this session.

This finishes our introduction to the Macintosh. Everything discussed above carries over to any Macintosh computer anywhere. We will learn more Macintosh basics as they become helpful, but now we're ready to become acquainted with Minitab.

You may have noticed, in particular, that we did not tell you how to shut down the Mac. That's because it should be left on most of the time; too much turning on and off can wear out the hard disk drive. The lab manager will shut the Macs down at the end of the day.

AN INTRODUCTION TO MINITAB

Minitab is a very powerful, yet user-friendly, data analysis application program. In this session, we will use it to examine test scores for an imaginary statistics class.

You can **launch Minitab** (or, as we like to say, "fire up Minitab") by following three steps:

1. Double-click on the hard disk drive icon (upper right of the desktop).

2. In the new window, double-click the Minitab folder icon.

3. In the next new window, double-click the Minitab icon.

The Minitab program icon is shown at the top center of the window in Figure 1.8.

You should see the Minitab logo, and after another moment you'll see a window, the **Data window**, named **Untitled Worksheet** and a new set of menus across the top of the screen. You might also notice the edges of another, inactive, window behind the Data window.

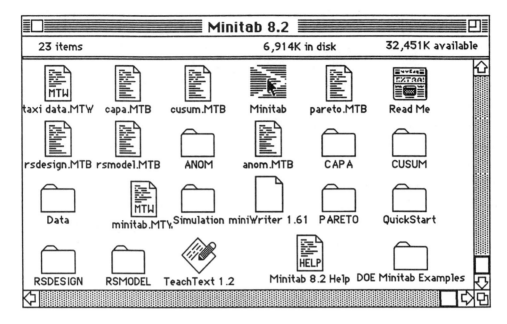

Figure 1.8 Minitab Program Icon in the Minitab Window

ENTERING DATA

Let's learn how to enter data by entering the information for our imaginary class. For each student in the imaginary class, we have a name, an Exam 1 score, and an Exam 2 score, shown in Table 1.1. Each row of the worksheet will correspond to the data for a student; we'll put students' names in the first column (**C1**), the Exam 1 scores in the second column (**C2**), and the Exam 2 scores in the third column (**C3**). Rows of a Minitab worksheet are called **observations**, and columns are called **variables**.

When we do work on data, we can refer to variables by their column numbers—C1, C2, and so on—or by more meaningful names we give the variables. Name the variables for this data set:

1. Figure 1.9 illustrates the Untitled worksheet. Click in the empty box immediately under **C1** and type **Student**. **Warning:** Do not type the variable names in the row labeled 1. You may have to retype the entire data set if you do.

2. Push <Tab>; this should move you one box to the right. Or, click in that box.

3. Type **Exam1**.

Table 1.1 Test Scores for an Imaginary Class

Student	Exam 1	Exam 2
Mozart	76	81
Sacajawea	88	91
Parks	72	85
Aristotle	87	88
Armstrong	80	86
Churchill	79	77
Dickinson	90	81
Tubman	77	82
Nightingale	92	81
DaVinci	90	88
Bronte	95	89
Anthony	71	79
King	93	90
Shakespeare	90	98
Gandhi	95	94
Jefferson	88	64
Renoir	75	88
Fossey	88	87
Curie	97	84
Earhart	77	85
Villa	82	97
Rudolph	48	56

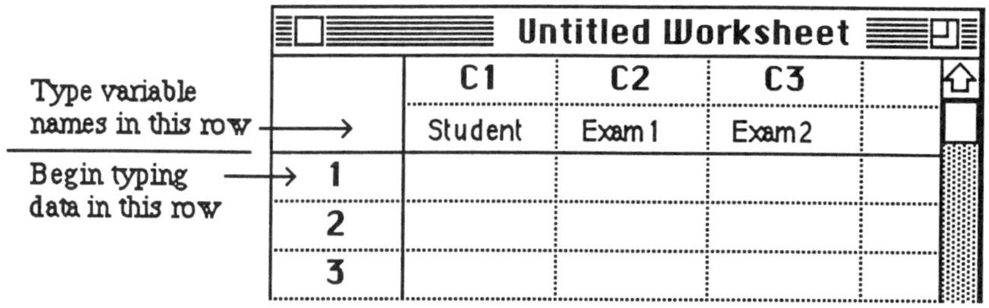

Type variable names in this row →

Begin typing data in this row →

	C1	C2	C3
	Student	Exam 1	Exam 2
1			
2			
3			

Figure 1.9 Untitled Worksheet

4. Tab again or click in the next box to the right, and then type **Exam2**.

Now begin to input the data:

1. Click in the first open box in column **C1** and then type **Mozart**.
2. Tab to the next cell and type **76**.
3. Tab to the next cell and type **81**.

Continue typing the names and test scores for all the students in this fashion. Figure 1.10 illustrates the Untitled worksheet after data for 14 students has been entered. If you are working in a team, let everyone have a turn. If you make a mistake, use <Delete> or click in the box containing the mistake and retype the entry. If you get to the bottom of the Data window, you can scroll the window down as explained in "Scrolling in Windows," below.

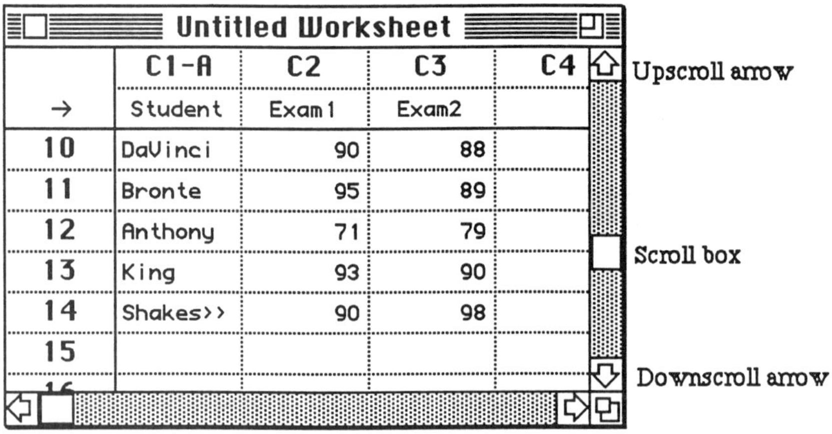

	C1-A	C2	C3	C4
	Student	Exam1	Exam2	
10	DaVinci	90	88	
11	Bronte	95	89	
12	Anthony	71	79	
13	King	93	90	
14	Shakes>>	90	98	
15				

Upscroll arrow

Scroll box

Downscroll arrow

Figure 1.10 Scrolling Tools

SCROLLING IN WINDOWS

Sometimes a Macintosh window gets too large to show onscreen in its entirety. For example, all the students' scores may not show in the Data window at once. As you're typing, the window should automatically give you more room, but early entries will then disappear from view. If you want to look at what's at the top of the Data window, you can **scroll up in the window** by clicking and holding on the **upscroll**

arrow that appears near the upper-right corner of the window (try it). Or, you may have to **scroll down in the window** by clicking and holding on the **downscroll arrow** at bottom right (try it). You may also drag the **scroll box**, which is between the scroll arrows, up or down to look at any part of the complete window. For example, drag the scroll box to the top of the scroll bar and then release the mouse button. Now you see the top of the window's contents. Now, drag the scroll box to the bottom of the scroll bar, and you will see the bottom of the window's contents. As soon as any Macintosh window gets too big to display in its entirety, scroll arrows will appear in the scroll bar. This is true for all Macintosh windows, not just in Minitab. There are also scroll arrows and a scroll box at the bottom of the Data window. These are used to scroll left and right. Figure 1.10 illustrates the scrolling tools in the Data window.

We entered the data one row at a time, using <Tab>. Data can also be entered one column at a time. To do it this way, you would click in the first cell of the column, type the entry (word or number), and then push <Return>. You will automatically move to the next cell down after you push <Return>.

Sometimes after inputting data, you will notice that some of the Minitab menu items are dimmed or in gray lettering, and that these items cannot be accessed. To correct this, click in any empty cell of the worksheet. These menu items should turn black again, and the items will become accessible.

After inputting data, double-check each number to make sure it is correct. *Always double-check data upon inputting it*, as input errors are very common, and if they are not corrected, many calculations will be in error, too. As the saying goes, "garbage in, garbage out." If it is not too disruptive, the best way to double-check data is to have someone else quietly read the data to you while you check it. If you find a mistake, click in that box and retype the value.

When you are sure the data set is correct, save and name the worksheet as follows:

1. Make sure your diskette is in the disk drive.

2. Under the **File** menu, select **Save Worksheet As**. A dialog box similar to Figure 1.11 will appear.

3. You should see your diskette's icon and name appear at the top right of the dialog box, signifying that the data will be saved to that diskette. If you don't see your diskette's icon, click the Drive button until you do.

4. Type **Grades** as the name for this worksheet.

16   SESSION ONE

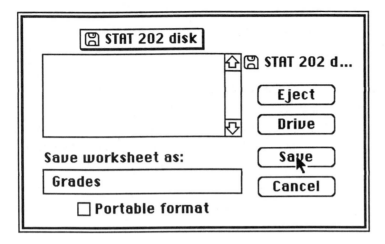

Figure 1.11 Save Worksheet As Dialog Box

5.   Click the **Save** button.

The title of the worksheet should now be **Grades.MTW**. Also, there should now be a file on your diskette with that name.

MODIFYING DATA: AN EXAMPLE

Minitab makes routine calculations and data sorting a snap. Let's use it to compute a mean score for each student, and then to sort the student records from greatest to least according to this mean score.

To compute the average exam score for each student, which is the average for each row in the worksheet:

1.   Click and hold on the **Calc** menu.

2.   While holding down the mouse button, move down the menu until **Functions and Statistics** is highlighted, but don't release the mouse button. Notice that another menu, a **submenu**, appears.

3.   Without releasing the mouse button, carefully move the mouse directly to the right until the pointer touches the submenu. If you slip, you'll have to go back to item 2 and try again.

AN INTRODUCTION TO THE MACINTOSH® AND TO MINITAB®   17

4. Move down this submenu until **Row Statistics** is highlighted. Then, release the mouse button, thereby selecting this item. A Row Statistics dialog box similar to Figure 1.12 will appear.

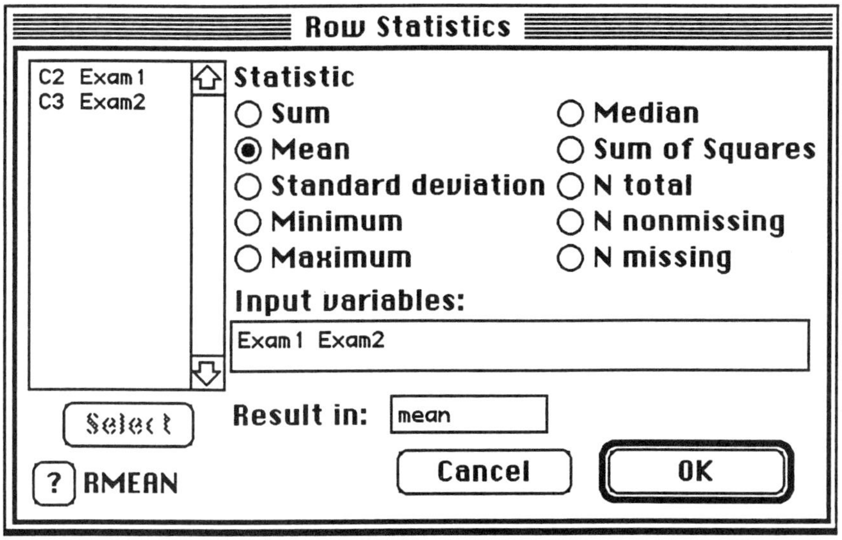

```
╔═══════════════════ Row Statistics ═══════════════════╗
│ C2 Exam1   Statistic                                 │
│ C3 Exam2    ○ Sum              ○ Median              │
│             ⦿ Mean             ○ Sum of Squares      │
│             ○ Standard deviation ○ N total           │
│             ○ Minimum          ○ N nonmissing        │
│             ○ Maximum          ○ N missing           │
│             Input variables:                         │
│             ┌──────────────────────────────────────┐ │
│             │ Exam1  Exam2                         │ │
│             └──────────────────────────────────────┘ │
│  [ Select ] Result in: [ mean    ]                   │
│  [?] RMEAN           [ Cancel ]   [[   OK   ]]       │
╚══════════════════════════════════════════════════════╝
```

Figure 1.12 Row Statistics Dialog Box

5. At the upper left of the dialog box is a list of the data set's numeric-variable columns (**C2, C3**) and their names. Click on **C 2** and then click the **Select** button. Now repeat this for **C3**. You have just selected these columns for the calculations. Several other ways to select variables will be demonstrated later.

6. Click in the circle to the left of the word **Mean**. You've just told Minitab that you want to compute the mean of the selected columns **C2** and **C3** for each row. Notice the other row calculations you could do.

7. Click in the box next to the words **Result in** and type **Mean**. You've just given a name to the first open column of the worksheet, column **C4**, which will be used to store the results of the calculation.

8. Click **OK**.

After some delay, the dialog box should close and you should see the worksheet, this time with a new column called Mean. The value of Mean in each row is the average of that student's Exam 1 and Exam 2 scores.

Now, it would be easy for you to see how the class as a whole did if the student records were sorted according to the overall mean score. To sort the data set by this variable:

1. Under the **Calc** menu, select **Sort**. A Sort dialog box similar to Figure 1.13 will appear.

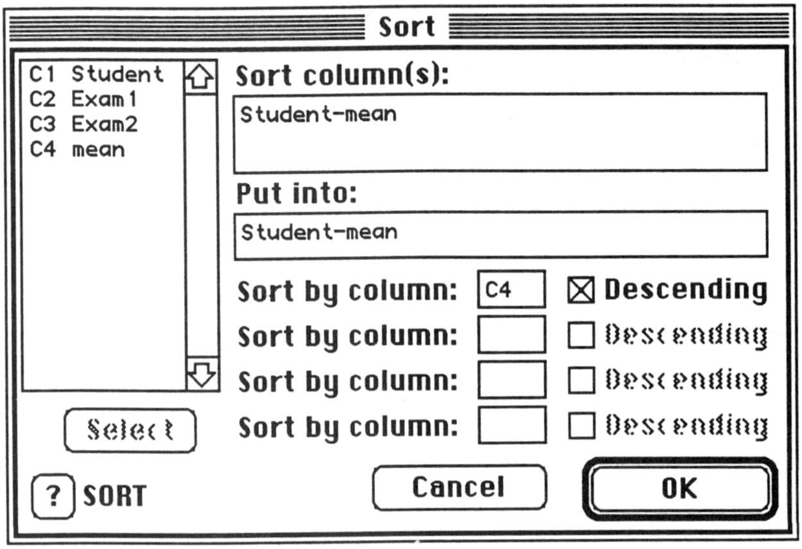

Figure 1.13 Sort Dialog Box

2. Click and hold on **C1** and then drag down to **C4**; this should highlight all four column names. Click the **Select** button. The column names should appear in the **Sort column**(s) box.

3. Click in the **Put into** box. Select each of the columns **C1–C4** again as in item 2. They should appear in the **Put into** box.

4. We want to sort according to the values in column **C4**, so in the first line that says **Sort by column**, click in the longer box and type **C4** or **Mean**.

5. Since we want the sort to be from highest mean score to lowest, click in the box to the left of **Descending**, next to the line where you just typed, and then click **OK**.

After some delay, you should see the Data window, but this time the students are listed in order of greatest to least mean test score. Who had the best average? Who had the worst? What grades would you give?

Right now, all the work you've done since the first time you saved the worksheet exists only on the screen in front of you. If the computer were to malfunction at this instant, or someone were to pull the plug, that work would be lost. You can save your changes to an existing worksheet at any time by choosing **Save Worksheet** under the **File** menu; try it now. This will replace the existing version of the worksheet on your diskette with the current version displayed on the screen.

The open-⌘ key, also called the **Command key**, is located at the lower left (and sometimes the lower right as well) on the keyboard. An easy way to save changes to an existing worksheet is by **simultaneously** pressing <Command> and S. This simultaneous pressing of two keys is abbreviated in this manual as **<Command>+S**. Try it. This is an example of a **keyboard command**. Many of the Macintosh menu items can be activated by using keyboard commands. If there is a keyboard equivalent for a menu item, the command sequence is shown next to the item in the menu.

If you want to keep both the old version of the worksheet and the new one, choose **Save Worksheet As** under the **File** menu, and give the new version a different name. Save your work often when working on any computer, because computers "crash" at the most annoying times. Get in the habit of saving your worksheet changes every few minutes.

GRAPHICAL DESCRIPTION OF A VARIABLE

It is difficult to see patterns in large collections of numbers just by looking at the numbers in tables. To better understand the class's distribution of mean exam scores, we could make a descriptive plot of some sort. Minitab offers four possibilities to graphically describe a single variable's distribution:

1. The dotplot
2. The stem-and-leaf plot (also called a stemplot)
3. The histogram
4. The boxplot (also called a schematic plot)

Perhaps you have learned about some of these in a statistics course or soon will. The first two are especially appropriate with small data sets like our class grade data.

We can make a **dotplot**, which is a "picture" of the mean exam score values, as follows:

1. Under the **Graph** menu, select **Dotplot**. A Dotplot dialog box similar to Figure 1.14 will appear.

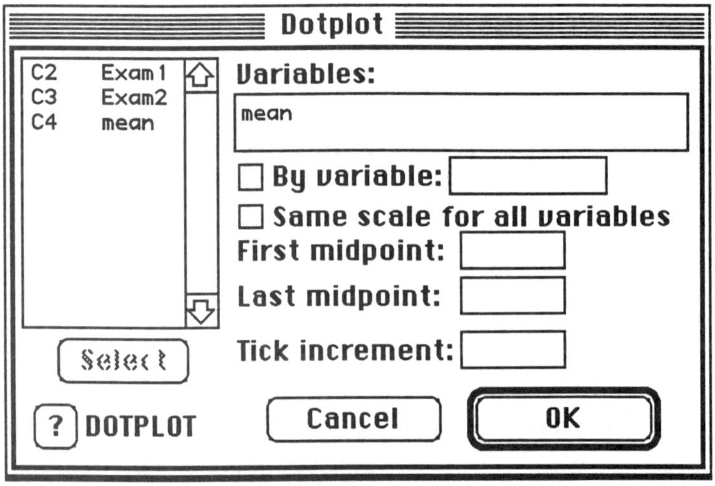

Figure 1.14 Dotplot Dialog Box

2. In the upper left of the dialog box, double-click on **Mean**. That variable name should now appear in the **Variables** box.
3. Click **OK**.

The **Session window**, which has been behind your Data window all along, now comes to the front (becomes **active**). After a few seconds, a dotplot of the mean exam-score data appears in the Session window.

THINK ABOUT THE DATA

The dotplot clearly shows most of the students' mean exam scores clustered between 75 and 100. However, there is one very low mean score, way down in the 50s. Many data analysts would call such an unusual value an **outlier**. One thing is sure; this student's mean score is not just in last place—it's way back. What's the problem? Is he or she a slow learner, or lazy? Don't fall into the trap of making quick and easy assumptions to explain outliers.

First, we need to find out which student is having trouble. Under the **Window** menu, select **Data**. The Data window will now appear. Look down the list of mean test scores to find the lowest one; you may have to scroll down the window. The lowest mean score belongs to Rudolph. Double-check the data in Table 1.1 to make sure Rudolph's Exam 1 and Exam 2 scores have been entered correctly.

As is often the case, careful consideration of an outlier leads to an important discovery: Rudolph is really a brilliant young student who has a testing disability. He has trouble holding the pencil to fill in your standardized test forms. All he needs is an alternative way to provide answers.

Another "picture" of the data that can be made is the stem-and-leaf plot. It's made just as the dotplot above except, in item 1, select **Stem-and-leaf** under the **Graph** menu. Make a stem-and-leaf plot of the mean score data now. This plot will also appear in the Session window. One advantage of the stem-and-leaf plot over the dotplot is that the former can show you the (rounded) value of each number. For example, in this plot we can see that the lowest mean exam score, rounded to the nearest whole number, is 52. The dotplot only showed us that it was a value in the 50s.

We're done modifying and inspecting the data, and in a minute we'll get a printout. First, let's look at some of the other windows.

THE FIVE MINITAB WINDOWS

Open the **Windows** menu. You will notice several items, corresponding to different types of windows that Minitab can display:

1. The **Data window** shows the worksheet in row and column form. From here, you can create new variables, modify old ones, sort data, subset the data, and much more.

2. The **Info window** gives summary information about a data set. Select that item now. You should see a list of columns, **C1–C4**, with names. Also, the "A" next to **C1** means that the variable Student is an **alpha variable**; its values are character strings. The other variables are **numeric variables**, and their values are numbers. Minitab treats these two types of variables differently at times. The Info window will not be extremely useful to us because we will use mostly small data sets, which we can inspect adequately in the Data window.

3. The **Graph window** will display certain graphs and plots that we will make, such as histograms and boxplots. You may not see it in the list of windows now, because we've not made any of those fancier graphs yet.

4. The **Session window** will be valuable to us. Select that item now under the Windows menu. Figure 1.15 shows a portion of a Session window for some example data. Besides the dotplot and stem-and-leaf plot, there are other lines beginning with "MTB>" or "SUBC>". The fact is, though every action we've carried out has been done by selecting menu items, each could also have been done by typing and entering commands in the Session window. Some types of work are much easier to do this way. For example, click just to the right of the last MTB>, which is called the **Minitab prompt**, and type **Note: This printout belongs to** _____ (fill in your name). Now, push <Return>. That was a **Minitab comment**. Any command beginning with **Note** is a Minitab comment. Next to the new Minitab prompt, type **Print C1-C4** and push <Return>. The computer responds to this command by printing the specified columns in the Session window. Once you gain enough familiarity with Minitab, typing commands in the Session window can be faster than selecting menu items. You can even write entire Minitab programs, called **macros**. Also, Minitab versions exist for PCs, microcomputers, workstations, and mainframes. These versions

often use commands instead of menus. The command version of Minitab existed well before the menu-driven version.

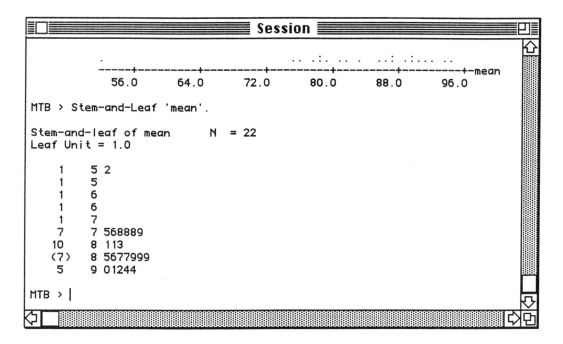

Figure 1.15 Session Window

5. The **History window** is like the Session window only less messy, and you can't enter any commands in the History window.

Next, we will obtain a printout of the Session window, so you have a record of the work you've done. It will also contain a printout of the final data set. The instructions below assume you will be printing on an ImageWriter® printer. If not, your instructor will explain any necessary modifications to the instructions, which will be very minor.

1. Make sure the Session window is active by selecting **Session** in the **Window** menu.

2. Under the **File** menu, select **Print Window**. A Print dialog box similar to Figure 1.16 will appear.

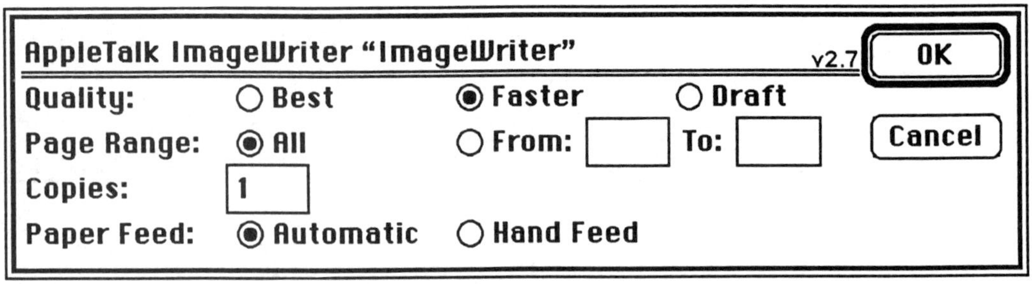

```
┌─────────────────────────────────────────────────────────────┐
│ AppleTalk ImageWriter "ImageWriter"           v2.7  [  OK  ]│
│ Quality:      ○ Best       ● Faster      ○ Draft            │
│ Page Range:   ● All        ○ From: [  ]  To: [  ] [Cancel]  │
│ Copies:       [ 1 ]                                         │
│ Paper Feed:   ● Automatic  ○ Hand Feed                      │
└─────────────────────────────────────────────────────────────┘
```

Figure 1.16 Print Dialog Box for ImageWriter Printer

3. Click in the circle next to the word **Faster**. We will usually print in Faster mode since it is fast yet gives a better quality printout than the Draft mode. Also, some Minitab output cannot be printed using the Draft mode.

4. Click **OK**.

The printer for the Macintosh should soon begin printing your Session window. When the printer is finished, *carefully* remove your output as follows:

1. On the printer control panel, push the **Select** button.

2. Push the **Form Feed** button.

3. Push the **Select** button again.

4. Gently tear off your hard copy, without yanking.

Make a printout of the Session window for each member of the team by repeating the steps above as many times as needed. Before printing, you may want to add a comment in the Session window showing the name of the person printing.

QUITTING MINITAB

Under the **File** menu, select **Quit**. Sometimes when you do this, Minitab will open a window asking whether you want to save changes you have made to the worksheet. If so, make your decision and click the appropriate button.

COPYING FILES BETWEEN DISKETTES

If you are sharing a computer, the last task for this session is to copy the worksheet **Grades.MTW** to each team member's diskette.

1. Close windows if necessary to get to the blank desktop. The icon for the diskette containing the file to be copied, the **source diskette**, should be at the upper right under the hard disk drive's icon.

2. Double-click on the source diskette's icon. In the newly opened window is an icon for each file stored on the diskette. One of these files should be the one you want to copy, in this case **Grades.MTW**. You may have to resize or scroll in the window to find it.

3. Push **<Command>+E**. This should eject the source diskette. The windows on the screen will change in appearance.

4. Insert the diskette that is to receive the file, the **destination diskette**. Its icon should appear under the source diskette's icon.

5. In the open window on the screen, click and hold on the icon of the file to be transferred, in this case **Grades.MTW**, and drag it across the screen onto the destination diskette's icon. When that icon darkens, release the mouse button.

6. The computer will eject the destination diskette and ask you to insert the source diskette. Do this. It will then eject that diskette and ask for the destination diskette. You may go through several iterations of this. Be patient and follow the computer prompts.

7. When finished, eject both diskettes by dragging their icons to the trash.

PARTING GLANCES

We have covered a lot of material about the Macintosh and Minitab. It may have been a bit of a blur. You'll find that, in time, working on these computers and with this software will not require a lot of "remembering" on your part. You'll develop instincts to know where to point and click in most situations.

More important, we found that statistics can be a tool to aid discovery. The simple process of detecting an outlier using a dotplot and then vigorously pursuing the question "Why is this score unusual?" averted a disaster. Rudolph, had he failed the course, may have become disillusioned with our educational system, dropped out, and turned to a life of crime. Instead, his academic career took off. He is now a leader, lighting the way for those with testing disabilities.

A more serious example is that inappropriate treatment of outliers led to a long delay in the discovery of the ozone hole above the South Pole. Satellite atmospheric monitoring instruments had been programmed to simply delete measurements that were unusually small or large—that is, to delete outliers—without investigation.

Name _____   Section _____   Session 1

SHORT ANSWER WRITING ASSIGNMENT

All answers should be complete sentences.

1. What feature of the dotplot led us to discover an outlier?

2. How did we discover that Rudolph was the outlier?

3. Give a real-world example (other than ignoring low scores in a classroom setting) where the thoughtless discarding of outliers caused the delay of an important discovery.

Provide a brief explanation for each of the following terms.

4. Icon

5. Desktop

6. Desk accessories

7. Resize box

8. Double-density diskette

9. Minitab observations

10. Session window

SESSION TWO
Variability and Heart Rates

INTRODUCTION

We do not always obtain the same results when we repeat an experiment. If it always came out the same way, it would be a pretty dull experiment. It is important to know not only the average outcome of an experiment, but also the amount and pattern of variability in the outcomes from repetition to repetition.

STATISTICAL CONCEPTS

Variability within and among individuals, effect of sample size, preliminary study, histogram, boxplot, numerical descriptive statistics.

MATERIALS NEEDED

For each team of two students: a Macintosh with the Calculator and Alarm Clock desk accessories, or a calculator and a watch with a second hand or a stopwatch.

THE SETTING

You are a member of a team conducting research in exercise physiology. Your team wants to determine the effects of an exercise program on subjects' heart rates. You have been assigned to conduct a **preliminary experiment** to determine how to measure an individual's heart rate and the amount and pattern of variability among individuals' heart rates. The latter would be necessary to decide how many subjects should be included in the full-scale experiment. Since we have no real subjects, each of you must in turn play the roles of researcher and subject. This would not be acceptable for an actual study!

BACKGROUND

Suppose that an expert has previously stated that counting heartbeats for 15 seconds and then multiplying by 4 to get beats per minute offers a good compromise between accuracy and convenience. In Phase 1, we investigate the consistency of this measurement strategy and compare it to a reasonable alternative. The question of how many subjects are necessary must always be answered before any serious study begins. To do this carefully, the amount and pattern of the measurement's variability among subjects are investigated in Phase 2.

THE EXPERIMENT, PHASE 1: MEASURING HEART RATES

STEP 1: DATA COLLECTION

An experiment must be precisely defined and carefully carried out to cut down on the extra variability that would be created if everyone did it differently. *Read this entire step before beginning data collection.*

Pair up and decide which member of the pair will first be the subject and which, the researcher. The subject will count his or her own heart rate. To measure your pulse, first turn your left wrist up. With the first two fingers of your right hand (fingers together), firmly press your left wrist just below the base of your thumb. You should feel a pulse. (Some individuals cannot feel wrist pulse. If this happens, try several places on your wrist, and if there is still no success, an alternative method is to press two fingers against your neck just to the right of center. We will ignore for simplicity the fact that the neck method is thought to give slightly higher heart rate readings.)

The Macintosh Alarm Clock will be used to time the counts. Select **Alarm Clock** under the menu. The researcher will keep time while the subject silently counts *without* watching the clock. The researcher should give the subject a silent signal to start counting when the seconds show 00, and another signal to stop when the seconds show 15. Write the heartbeat count in the top blank of the first column of Table 2.1 in the **subject's** manual.

That was one replication, or rep, of the experiment to measure heart rate. To study variability of this measurement technique, we use eight reps. *Don't make a conscious effort to make your counts agree with each other.* Scientists call that fudging the data; it's one of the biggest causes of faulty research conclusions. Some heart-rate counts are large and some counts are not so large; that's variability and is what we

Table 2.1 Heartbeats/Minute Using 15-Second Counts

15-Second Count	Multiplier	x = Rate15	x^2
	* 4 =		
	* 4 =		
	* 4 =		
	* 4 =		
	* 4 =		
	* 4 =		
	* 4 =		
	* 4 =		
Column sum	----------	Σx =	Σx^2 =

Sample size = n	8
Mean = $\bar{x} = (\Sigma x)/n$	
Sample variance = $s_x^2 = [\Sigma x^2 - (\Sigma x)^2/n]/(n-1)$	
Standard deviation = $s_x = \sqrt{s_x^2}$	

are studying. Repeat the heart-rate counting experiment seven more times. Don't get in a hurry. Record each count in the first column of Table 2.1 in the subject's manual.

Now exchange roles of subject and researcher. When each subject has finished recording eight 15-second counts, multiply these counts by four to obtain beats per minute. Write the results in the third column of Table 2.1.

Now, repeat the entire data collection routine for each subject using 30-second counts rather than 15-second counts; do this eight times for each subject. Write the 30-second counts in the first column of Table 2.2. After all data collection is finished, multiply these counts by two to obtain beats per minute. Write these heart rates in the third column of Table 2.2.

Table 2.2 Heartbeats/Minute Using 30-Second Counts

30-Second Count	Multiplier	y = Rate30	y^2
	* 2 =		
	* 2 =		
	* 2 =		
	* 2 =		
	* 2 =		
	* 2 =		
	* 2 =		
	* 2 =		
Column sum	----------	$\Sigma y =$	$\Sigma y^2 =$

Sample size = n	8
Mean = $\bar{y} = (\Sigma y)/n$	
Sample variance = $s_y^2 = [\Sigma y^2 - (\Sigma y)^2/n\,]/(n-1)$	
Standard deviation = $s_y = \sqrt{s_y^2}$	

STEP 2: MANUAL DATA ANALYSIS

Compute the mean, variance, and standard deviation of your 15-second heart-rate values in Table 2.1 using a hand calculator or the **Calculator** desk accessory in the menu. Table 2.1 is laid out to help you with the calculations. The formula for the variance is a very complicated, highly abbreviated set of instructions. You may prefer following the step-by-step guide. Record the results of each step in the corresponding blank at the right of the guide.

1. Sum the data values (column 3 sum). $\Rightarrow$ 1 _____
2. Sum the squared values (column 4 sum). $\Rightarrow$ 2 _____
3. Square the result shown on line 1. $\Rightarrow$ 3 _____

4. Divide the result on line 3 by n, 8 in this case. $\Rightarrow$ 4 _____

5. Subtract the result on line 4 from the result on line 2. $\Rightarrow$ 5 _____

   (If the result is negative, you've made a mistake.)

6. Divide the result on line 5 by $(n - 1)$, 7 in this case. $\Rightarrow$ 6 _____

If these steps have been carried out carefully, the value on line 6 is the sample variance of the values in column 3 of Table 2.1. You'll be able to check this calculation later using Minitab. Write your results for the mean, variance, and standard deviation in Table 2.1. When you are finished with the calculations for the 15-second values, do the calculations for the 30-second heart-rate values in column 3 of Table 2.2 and write the results in Table 2.2.

STEP 3: DATA ENTRY

Manual calculations are a lot of work! Though it is useful to know how to do these basic calculations by hand, we will use Minitab to do routine calculations in the blink of an eye. Once the data is entered into a Minitab worksheet, we can easily do these sorts of calculations and much more. Let's create a worksheet for the Phase 1 heart-rate data as follows. If you are sharing a computer, one partner should complete all of the data entry section and then change places and let the other partner complete the step using his or her heart-rate data. Help each other. Prior to entering your data, make sure *your* diskette is in the disk drive. Then, launch Minitab (if you don't remember how, see Session 1). After a short delay an untitled worksheet will appear. Name the first two columns **Rate15** and **Rate30**. Carefully enter your heart-rate data from column 3 of Table 2.1 for the Rate15 variable in the worksheet. Then, enter the values from column 3 of Table 2.2 for the Rate30 values. Figure 2.1 illustrates the variable names in the Untitled worksheet.

Double-check the inputted data before going any further. Have your partner help by reading the values to you. When you are sure the data is correct, save the worksheet onto your diskette with the name **heartrates(mine)** by selecting **Save Worksheet As** under the **File** menu. Figure 2.2 shows the Save Worksheet As dialog box. When you have finished, the title of the worksheet should change to **heartrates(mine).MTW**. Also, there will be a file with that name on your diskette. Quit Minitab by selecting **Quit** under the **File** menu, and eject your diskette by dragging it to the trash can.

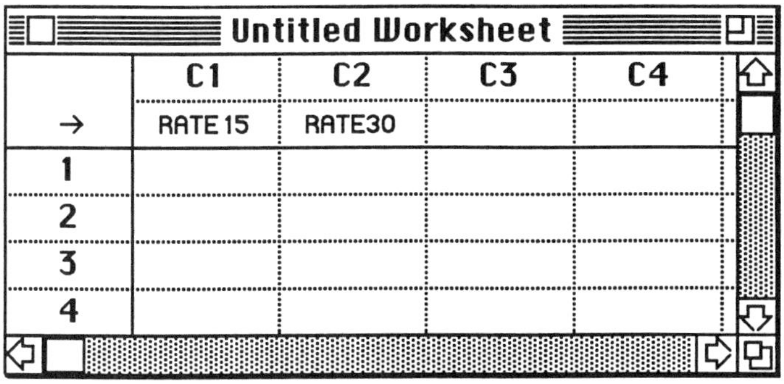

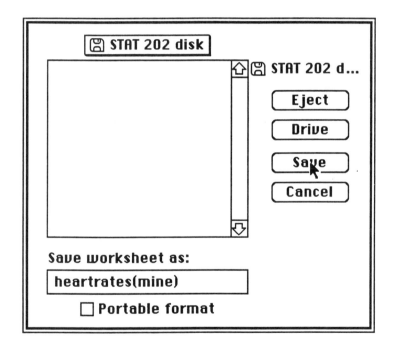

Figure 2.1 Variable Names in the Untitled Worksheet

Figure 2.2 Save Worksheet As Dialog Box

STEP 4: COMPUTER DATA ANALYSIS

You should start this step with a clean desktop on the computer and no diskette in the disk drive. You can open an existing Minitab worksheet at any time once a worksheet has been created and saved on your diskette:

1. Insert your diskette into the disk drive.

2. Double-click on your diskette's icon. A window showing the contents of your diskette will open.

3. Double-clicking on a Minitab worksheet icon launches Minitab and also opens that worksheet. There should be an icon on your diskette labeled **heartrates(mine).MTW**. Double-click on this icon.

You should see the Minitab logo for a moment, and then the **heartrates(mine).MTW** worksheet will open and you will see the data you entered a few minutes ago.

We can ask Minitab to compute descriptive statistics for a variable as follows:

1. Under the **Stat** menu, click and hold on **Basic Statistics**. Move the mouse directly to the right and select **Descriptive Statistics** from the submenu. A Descriptive Statistics dialog box similar to Figure 2.3 will appear.

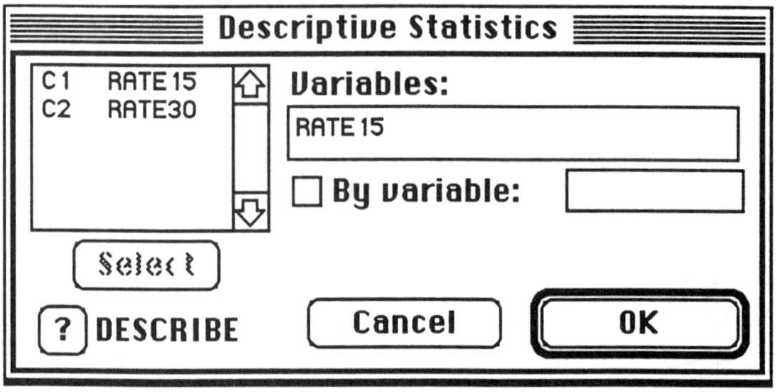

Figure 2.3 Descriptive Statistics Dialog Box

2. Click in the box under **Variables** and then double-click on **Rate15**.

3. Click **OK**.

After a short delay, output will appear in the Session window. Figure 2.4 shows the Session window with some example data. The descriptive statistics are shown under the headings "N," "MEAN," "MEDIAN," "TRMEAN," "STDEV," and "SEMEAN." N is the number of values for the variable. The median is a value that splits the ordered-least-to-greatest list of values in half. We will not concern ourselves with TRMEAN and SEMEAN here, but the other computed values should match your manual calculations from Table 2.1 for the mean and standard deviation of the Rate15 values. Check these now. There may be some small differences, especially in the standard deviation, due to roundoff error in your manual calculations. If there are any serious discrepancies, you have made a mistake, either in the manual calculations or while inputting the data.

The second set of values shown in the Session window is headed by the words *MIN, MAX, Q1*, and *Q3*. The values below the headings "MIN" and "MAX" are the smallest and largest of the Rate15 values. Q1 is the first quartile (25th percentile) of the Rate15 values, and Q3 is the 3rd quartile (75th percentile) of the Rate15 values. Of course, the median is the second quartile (50th percentile) of the values.

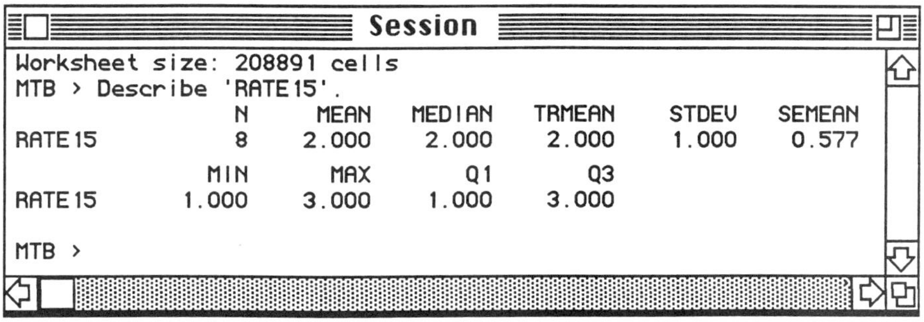

```
================================ Session ================================
Worksheet size: 208891 cells
MTB > Describe 'RATE15'.
             N      MEAN    MEDIAN    TRMEAN     STDEV    SEMEAN
RATE15       8     2.000     2.000     2.000     1.000     0.577

           MIN       MAX        Q1        Q3
RATE15   1.000     3.000     1.000     3.000

MTB >
```

Figure 2.4 Session Window with Example Data

To get descriptive statistics for the 30-second counts (Rate30), we could complete the above items 1–3, but it is much simpler to use the **Descriptive Statistics command** in the Session window:

1. Click to the right of **MTB>** and type **Describe 'Rate30'**.

2. Push **<Return>**.

Note that, in item 1, we enclosed the variable name in single quotation marks. We also could have referred to the variable by its column number without quotation

Table 2.3 Comparisons of Heart-Rate Descriptive Statistics ($n = 8$)

Statistic	Rate15	Rate30
Mean		
Standard deviation		
Range		

marks by typing, **Describe C2**. After a short delay, the results of descriptive calculations on the values of Rate30 should appear in the Session window. Check these against your manual calculations in Table 2.2.

Enter the values of the computer-generated descriptive statistics shown in the Session window for Rate15 and Rate30 in the appropriate cells in Table 2.3. You can compute the **Range** = MAX - MIN of the Rate15 values by hand, and similarly for the Rate30 values. You may have to scroll up or down in the Session window to find everything you need. This concludes our calculations with the data for Phase 1. Make sure you have all the output you need, and then quit Minitab by selecting **Quit** from the **File** menu.

When the entire class has completed Table 2.3, the instructor will ask each student to read his or her results. Record the descriptive statistics for each student in your Table 2.4. After all of the data is recorded, the instructor will lead a class discussion concerning which method of heart-rate measurement is more reliable: a 15-second count multiplied by four or a 30-second count multiplied by two.

THE EXPERIMENT, PHASE 2: VARIABILITY AMONG SUBJECTS

In this phase we look at variability in heart rates among individuals. If there is a great deal of variability in heart rates among individuals, we would need many subjects to adequately test the effects of an exercise treatment, or we would need a more complicated experiment.

Table 2.4 Summary Statistics for Each Student's Heart-Rate Measurements

Student	Rate15			Rate30		
	Mean	Standard Deviation	Range	Mean	Standard Deviation	Range
1						
2						
3						
4						
5						
6						
7						
8						
9						
10						
11						
12						
13						
14						
15						
16						
17						
18						
19						
20						
21						
22						
23						
24						

STEP 1: DATA ENTRY

Put a diskette in the disk drive and launch Minitab. An untitled worksheet will appear. Name the first variable **Mean**. Enter each student's mean Rate30 value from column 5 of Table 2.4 into this column of the worksheet. After entering the data, double-check the values in the worksheet against the entries in Table 2.4. When you are sure your data is correct, save your worksheet onto your diskette as **heartrates(class)** by selecting **Save Worksheet As** under the **File** menu.

STEP 2: NUMERICAL AND GRAPHICAL DESCRIPTION OF DATA

Let's use Minitab to compute descriptive statistics on the entire class's values for Mean, just as we did for everyone's Rate15 and Rate30 values in Phase 1.

1. Under the **Stat** menu, click and hold on **Basic Statistics** and then select **Descriptive Statistics** from the submenu. The Descriptive Statistics dialog box will appear.

2. Click in the box under **Variables** and then double-click on **Mean**.

3. Click **OK**.

The summary statistics for the set of students' mean heart rates should appear in the Session window.

We can also use Minitab to graphically describe this data. For example, we could make a dotplot or a stem-and-leaf plot, as in Session 1. Instead, let's try a histogram:

1. Under the **Graph** menu, select **Histogram**. A Histogram dialog box similar to Figure 2.5 will appear.

2. Click in the box next to **Variables** and then double-click on **Mean**.

3. Click **OK**.

A GHistogram window will open and a histogram of the students' mean heart rates will appear. The histogram shows how many values fall into each of several class intervals along the number line.

We'll print the histogram out in a minute, but first let's make a boxplot of the data:

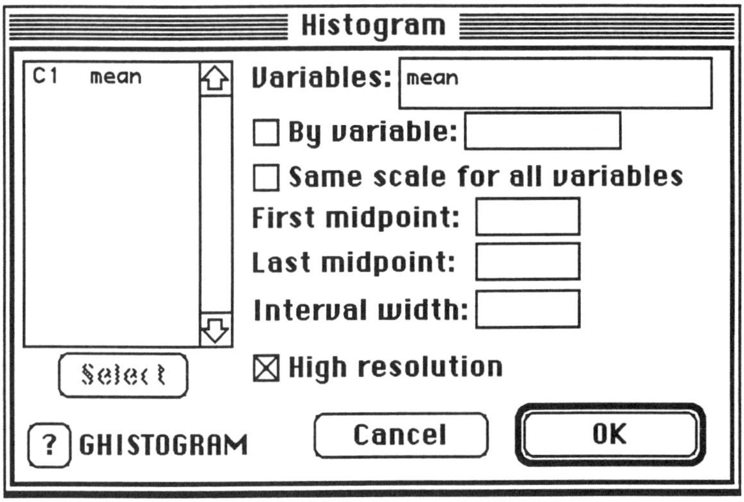

Figure 2.5 Histogram Dialog Box

1. Under the **Graph** menu, select **Boxplot**. The Boxplot dialog box will appear.

2. Click in the box next to **Variables** and then double-click on **Mean**.

3. Click **OK**.

A GBoxPlot window will open and a boxplot of the data will appear. The boxplot is a schematic plot showing the location of MIN, MAX, Q1, MEDIAN, and Q3.

STEP 3: PRINT THE RESULTS

In this step, be patient with others who may be sharing the printer with you. We will make copies of the printout for each team member. First, print the Session window:

1. Under the **Window** menu, select **Session**.

2. Next to the Minitab prompt, **MTB>**, type **Note: This printout belongs to** and then type your name(s).

3. Under the **File** menu, select **Print Window**.

4. Click in the box next to **Copies** and type the required number of printout copies. You may also need to click a button to indicate the desired print quality. Figure

2.6 illustrates the **Print Window** dialog box for an ImageWriter printer using the Faster print quality with 2 copies.

5. Click **OK**.

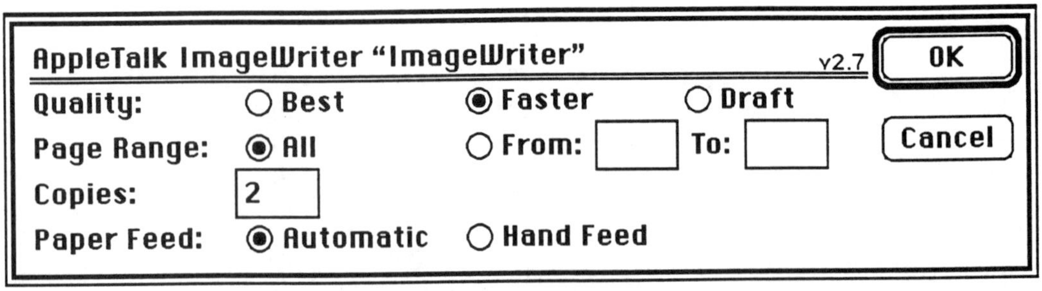

```
AppleTalk ImageWriter "ImageWriter"                v2.7    [  OK  ]
Quality:        ○ Best          ● Faster      ○ Draft
Page Range:     ● All           ○ From: [   ]  To: [   ]  [ Cancel ]
Copies:         [2]
Paper Feed:     ● Automatic     ○ Hand Feed
```

Figure 2.6 Print Window Dialog Box for ImageWriter Printer

After a short delay, the printer for the Mac should begin your printout. When it is finished, you can then print the histogram or boxplot. Note that the histogram and boxplot are high-resolution graphics and cannot be printed by an ImageWriter printer in the Draft mode.

1. Under the **Window** menu, select the desired graph window. Window options are listed at the bottom of the menu.

2. Repeat items 3–5 from the above print instructions.

Print as many copies of the histogram and boxplot as your team needs.

When all copies of all of your team's output have printed, carefully remove the output from the printer as described at the end of Session 1 and quit Minitab. To do this, select **Quit** under the **File** menu, or try the keyboard shortcut **<Command>+Q**.

If you are sharing a computer, the last task for this session is to copy the worksheet **heartrates(class).MTW** to each team member's diskette. Refer to "Copying Files Between Diskettes" section in Session 1, making modifications to worksheet names as needed.

PARTING GLANCES

The basic measurement for this experiment was a count of a person's heartbeat for a fixed length of time. How long should we count? Theoretically, for any person, longer lengths of time should give about the same average rate per minute as shorter lengths, but there should be less variability in a rate based on a longer time. This is because we can think of the beats per minute as an average, and averages over 30 seconds should be more reliable than averages over 15 seconds. Phase 1 of this session was meant to address questions concerning **variability within subjects**: Which of the two choices of basic measurement has less variability?

In Phase 2, we adopted the average of eight 30-second rate-per-minute counts as the basic measurement and investigated the **variability among subjects**. It is important in a preliminary study to measure the variability in the basic measurement among subjects in order to make a sound decision on the sample size (number of subjects) for the full-scale study. Why? Essentially for the same reason as above: The effects of the exercise treatment will be reported as an average of some sort over measurements made on the subjects in the full-scale study. An average over 200 subjects will be more reliable (have less variability) than an average over 100 subjects, but how precise will an average over 100 subjects be? How about an average over 200 subjects? Is the precision gained in using 200 subjects worth the extra expense and trouble? Answering questions like these is among the most important responsibilities of a professional statistician. All too often researchers dive headlong into years and thousands or hundreds of thousands of dollars' worth of data collection only to discover afterward that their basic measurement was too imprecise, or their choice of sample size too small, to yield reliable results.

EXTENDED WRITING ASSIGNMENT

Refer to Appendix 1, "Technical Report Writing," and Appendix 2, "Technical Report Writing Checklist," for guidance on format and style for your report.

A federal health agency has commissioned you to carry out today's experiment. Write a report to this agency that includes:

1. A description of the experiment, including how the data was collected.

2. A summary, in your expert opinion, of the conclusions drawn from Phase 1 of the experiment concerning the question of which heart-rate measure, 15 seconds or 30 seconds, seems more reliable.

3. Presentation of the class's mean Rate30 data graphically and in a table.

4. Description of the general shape of the distribution of the class's mean Rate30 data. Is there more variability among repeated measures of Rate30 on yourself or among mean Rate30 measures of all the class members?

Name   Section   Session 2

SHORT ANSWER WRITING ASSIGNMENT

All answers should be complete sentences. Include copies of Tables 2.3 and 2.4 and the histogram of the class's mean Rate30 data, appropriately titled and labeled, with this assignment.

1. Report the sample means for your Rate15 and Rate30 values. Comment on any difference that you find between these sample means.

2. Report the sample standard deviations for your Rate15 and Rate30 values. Comment on any difference that you find between these sample standard deviations.

3. Report the sample ranges for your Rate15 and Rate30 values. Comment on any difference that you find between these sample ranges.

4. From the data in Table 2.4, how many students had a larger mean for their 15-second heart-rate measure, Rate15, than for their 30-second heart-rate measure, Rate30? What does this tell you about the two heart-rate measures for the class as a whole?

5. From the data in Table 2.4, how many students had a larger standard deviation for their 15-second heart-rate measure, Rate15, than for their 30-second heart-rate measure, Rate30? What does this tell you about the two heart-rate measures for the class as a whole?

6. From the data in Table 2.4, how many students had a larger range for their 15-second heart-rate measure, Rate15, than for their 30-second heart-rate measure, Rate30? What does this tell you about the two heart-rate measures for the class as a whole?

7. Circle each of the following items that describes the class's heart-rate counts as depicted in your histogram.

   Approximately symmetric

   Skewed left

   Skewed right

   Bimodal

   Has outliers

SESSION THREE
Author, Author

INTRODUCTION

It has become standard to attempt to identify the author of an anonymously published literary work by analyzing quantitative characteristics of the writing style in the work, such as average sentence length and rates of occurrence of certain words, punctuation, or phrases. These results are then compared to similar analyses on documents by authors who are considered candidates for the "mystery work." In effect, a statistical fingerprint is created for each of several candidate authors, and the anonymously written document is "dusted for prints."

STATISTICAL CONCEPTS

Classification, by-treatment dotplots, scatter plots.

MATERIALS NEEDED

For each team, a straightedge.

THE SETTING

It is the year 2109 and this text is now in its 16th edition. Its authors retired as billionaires and are now very famous and very dead. A session of the 1st edition, cut before publication, has recently been discovered in a discarded file cabinet. This session's authorship is being furiously debated by historians all over the world. Your job is to find quantitative clues as to the authorship by performing a quantitative style analysis on the disputed session and on the sessions whose authorship is known.

BACKGROUND

Though no author writes in exactly the same style from day to day, there are uniquely individual patterns in the frequency of use of common words and phrases and in other quantifiable characteristics such as sentence length. It would not be impossible for an author's style to change radically from one work to another, but it is improbable because writing habits are too deep-rooted. In this session we informally examine patterns in the variability of some writing style characteristics. More formal analyses, using the theory of probability, are possible. These analyses are known as **classification** or **discriminant analyses**.

THE EXPERIMENT

STEP 1: STUDY DESIGN

Each team will consist of four students, charged with trying to find evidence as to the authorship of the disputed session. A portion of the disputed session is reproduced in Figure 3.1. The three authors of this manual are referred to as authors 1, 2, and 3 but are not necessarily numbered in the order shown on the textbook cover. The authorship (1, 2, or 3) of each of sessions 1 through 17 is shown in the second column of Table 3.1. For now, the unknown author of the disputed session will be referred to as author 0.

Your data will be gathered as follows: The first page of each session will serve as a sample of its author's style and Figure 3.1 will be the sample from the disputed session. The number of words on each of these pages will be supplied by your instructor. Fill in the third column of Table 3.1 with these values. Your team will count four quantities on each sample. The number of sentences on each page will be counted, as well as the number of occurrences of three additional discriminators that your team will choose. The best discriminators are common words or punctuation marks used often in nearly any document. Typical choices for discriminating words are *a, and, the, of, to, in, be, that,* and *it*.

Of course, there are many words your team may choose for discriminators. If you choose words other than those listed above, be sure to choose frequently used ones. If you choose to use punctuation marks as discriminators, choose commonly used marks such as the comma, period, question mark, or semicolon. Also, avoid words whose use is topic driven; that is, words that would be used often by any author when writing on a specific topic. As an example, any of the three candidate authors would use the word

Session 18 Breaking Strength Experiment

Introduction
Measuring the **breaking strength** of a particular fiber or object has important applications in many areas. Civil engineers use truck-size instruments to measure the maximum load that pavement can withstand, and this information is then used toward the development of stronger materials for road construction. Stress/strength measurements are important to those working with any type of fiber. Examples include materials-science researchers, who use microscopic instrumentation to measure the strength of carbon fibers, or monofilament companies that, in an effort to maintain quality, routinely measure the strength of their products.

Statistical Concepts
Numerical summary statistics, boxplot, histogram, one-sample t-test and confidence intervals.

Materials Needed
For each team, force gauge, fishing line, kite string, wooden dowel about 1 foot long, and scissors.

The Setting
During the past couple of weeks, you have had your share of bad luck when it comes to leisure activities! Just last Saturday you were out at the local park flying your brand new Deluxe Super Hero kite. The wind was strong and the kite was soaring. Then, *pop*! Much to your dismay and disbelief, the kite string had broken and your kite was speeding upward to meet its master.

And then came this past Sunday's incident. It was a sunny afternoon perfect for a little fishing on the river. While you sat on the riverbank and pondered the statistics homework due on Monday, your fishing rod was nearly jerked from your grip. You had hooked a big one! You began to reel the monster in and could actually see it—a hulking 8- or 9-pound striped bass. But your dreams came crashing to a halt with the sound of a dull snap. Your fishing line had broken.

Background

Figure 3.1 First Page of the Disputed Manuscript

Table 3.1 Quantitative Writing Style Data

Session	Author	Words	Sentences			
1	3					
2	3					
3	3					
4	1					
5	3					
6	1					
7	2					
8	2					
9	2					
10	2					
11	2					
12	1					
13	1					
14	3					
15	3					
16	2					
17	1					
Disputed	0					

Macintosh often if writing Session 1, so *Macintosh* could be considered a topic-driven word and therefore not a good choice for a discriminator.

Whatever your team chooses for its discriminators, write appropriate headings in the last three columns of Table 3.1. For the rest of this discussion, the discriminators are referred to as discriminator 1, discriminator 2, and discriminator 3.

STEP 2: GATHERING THE DATA

Each member of the team should be a count specialist for one item. One member will count sentences; another, occurrences of discriminator 1; and so on. Before you begin counting, the team may need to make some operational definitions. For example, should you count sentences that end on the next page? Does the period in *1.* count as a period? How you settle these little dilemmas is less important than consistency in applying your decisions. This is one reason each team member will be a specialist in counting one item. Try to foresee as many problems as you can before counting and discuss them in advance. Write your operational definitions in the space below.

When your team feels prepared to count, proceed carefully, with each member filling in his or her column of Table 3.1. Don't hurry—the quality of the final analysis will be hurt by sloppily collected data. One interesting counting strategy is to count up from the bottom of the page to avoid the tendency to read as you count. When all have finished, exchange information until every team member's Table 3.1 is filled in.

STEP 3: DATA ANALYSIS

Work alone or in teams of two. Choose a Macintosh and launch Minitab. In the newly opened worksheet, type the variable names **Session**, **Author**, **Words**, and **Sents** in the boxes immediately below **C1**, **C2**, **C3**, and **C4**. We can't use *Sentences* as a variable name because the limit for Minitab variable names is eight characters. Also, type descriptive variable names for your team's three discriminators. Remember the eight-character limit and that the first character must be a letter. Don't use punctuation marks in variable names. In the discussion that follows, the names Disc1, Disc2, and

Disc3 appear for your team's discriminators, but you should create more descriptive names than these. Figure 3.2 illustrates a portion of the Untitled worksheet.

	C1	C2	C3	C4
→	SESSION	AUTHOR	WORDS	SENTS
1				

Figure 3.2 Untitled Worksheet

When you have named all the variables, input the data from Table 3.1 in the appropriate columns. *Be sure to use numbers (1, 2, 3, and 0), not letters (A, B, etc.), for the variable Author.* After you have entered the data and double-checked it for accuracy, save the data set onto your diskette as **Author.Author**.

We will now create a new variable, **Slength**, whose value for each session will be the number of words divided by the number of sentences, or the average sentence length.

1. Under the **Calc** menu, click and hold on **Functions and Statistics**. Move the mouse directly to the right and select **General Expressions** from the submenu. A General Expressions dialog box similar to Figure 3.3 will appear.

2. Click in the box to the right of **New/modified variable** and type **Slength**.

3. Click in the box under **Expression** and type **'Words'/'Sents'**. Alternately, you could type **C3/C4**.

4. Click **OK**.

The new variable **Slength** should appear in column C8. You may have to scroll the worksheet to the right to see it.

We need to adjust our discriminator counts in columns C5–C7 of the worksheet for the number of words on the page. A page with 500 words will naturally tend to have more occurrences of any given discriminator than a page with 300 words. We correct for this by creating new variables to reflect the **rate of occurrence per 100 words** of our discriminators. If we are counting occurrences of the word *the*, the rate per 100 words for a page is

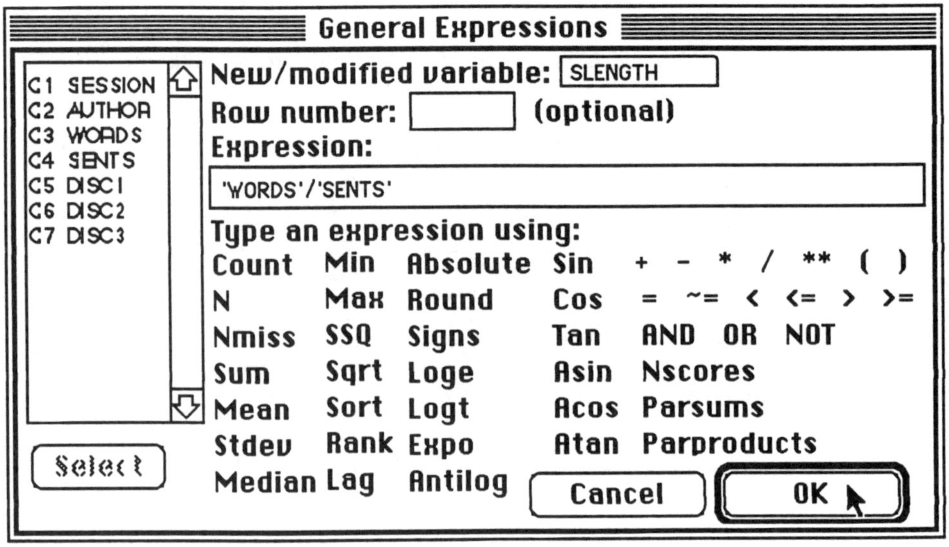

Figure 3.3 General Expressions Dialog Box

$$\text{rate of } the \text{ per 100 words} = \frac{100 * (\text{count of } the \text{ on the page})}{\text{word count on the page}}$$

Note that the rate of occurrence of a discriminator per 100 words will not necessarily be a whole number. We should not be concerned to hear that an author uses *the* at a rate of 12.6 times per 100 words, any more than we should be concerned to hear that there are, on average, 1.5 children per U.S. household. If we understand the definition of a rate, there is no logical dilemma.

The counts for your first discriminator should be in column C5. To calculate the rates per 100 words for your first discriminator:

1. Under the **Calc** menu, click and hold on **Functions and Statistics** and then select **General Expressions** from the submenu. A General Expressions dialog box will appear.

2. Click in the box next to **New/modified variable** and type a well-chosen variable name. For example, if you are calculating the rate of occurrence of *the*, you might call the new variable *Ratethe*. Remember, you must begin the variable name with a letter and use no more than eight characters. In illustrations to come, these rate variables appear as Rate1, Rate2, and Rate3, but you should use more descriptive variable names.

3. Click in the box under **Expression** and type **100*C5/C3**.

4. Click **OK**.

After a short delay, the new variable should appear in column C9. You may have to scroll the worksheet to the right to see it.

Repeat the above steps for your team's other discriminators, substituting your well-chosen variable names in item 2 and **100*C6/C3** or **100*C7/C3** in item 3. The new variables should appear in columns C10 and C11. After you have created the rate variables, save the modified worksheet by selecting **Save** under the **File** menu.

To statistically fingerprint our authors, we will begin with dotplots done separately for each author and laid out side by side. In this role, the identifier variable **Author** is sometimes called a **by-variable** or **subscript**. Before we can do these analyses by author, we have to sort the data set to put rows from the same author together in the worksheet:

1. Under the **Calc** menu, select **Sort**. A Sort dialog box similar to Figure 3.4 will appear.

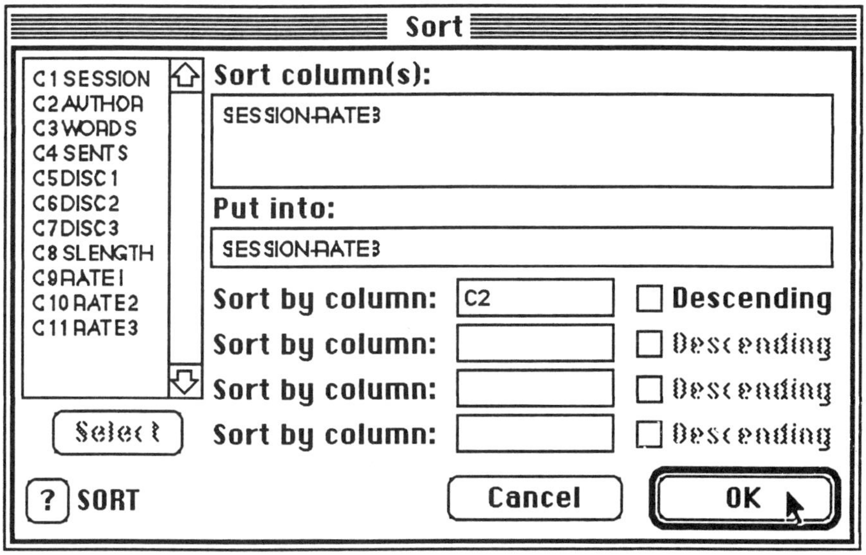

Figure 3.4 Sort Dialog Box

2. Click and hold on **C1** and then drag down to **C11**. This should highlight all 11 column names. Then click **Select**.

3. Click in the **Put into** box. Select the columns **C1–C11** again.
4. Click in the box after the first **Sort by column** and type **C2**.
5. Click **OK**.

To make side-by-side dotplots of mean sentence length for Authors 1, 2, 3, and 0:

1. Under the **Graph** menu, select **Dotplot.** A Dotplot dialog box similar to Figure 3.5 will appear.

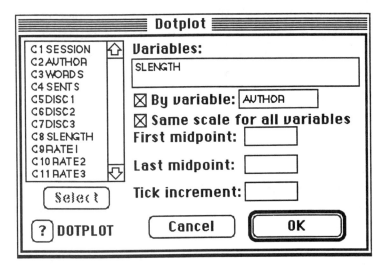

Figure 3.5 Dotplot Dialog Box

2. Click in the box under **Variables** and then double-click on **Slength**.
3. Click in the box to the left of **By variable**.
4. Click in the box to the right of **By variable** and then double-click on **Author**.
5. Click in the box to the left of **Same scale for all variables**.
6. Click **OK**.

After a short delay, four dotplots of the mean sentence length with the same scale should appear in the Session window. These dotplots can be used to compare the authors on the basis of their mean sentence-length data. Of course, the dotplot

Table 3.2 Example Mean Sentence-Length Data

Author 1:	8	9	10	11	12
Author 2:	9	10	10	10	11
Author 3:	6	8	10	12	14
Author 0:	13				

for Author 0 only has 1 point; it is just there to show whether that point fits in the pattern of values for the known authors.

Repeat the above dotplot construction for each of your team's discriminator rates. You can use the same instructions, except in item 2 double-click on one of your other discriminator rates (*don't use the raw counts*) instead of the variable name **Slength**. You'll have to do the dotplot construction a total of four times.

Before continuing, print your dotplots by selecting **Print Window** under the **File** menu. This printing will probably take a little while. Be careful not to pick up somebody else's output. While you are waiting for your output, reread the next step.

STEP 4: THINK ABOUT THE DATA

Use your dotplots to gain clues as to which author is Author 0, the unknown author of this session. For example, if the mean sentence length for Author 0 is quite long and only Author 2 has a comparable mean sentence length, that points to Author 2 as the mystery author. Or, perhaps both Author 1 and Author 2 have rates of occurrence for discriminator 1 comparable to that of Author 0, but Author 3's rates are very different from the Author 0 rate; that suggests Author 3 didn't write the disputed session, which is at least a partial conclusion.

We should consider not only the typical overall values of a discriminator, but also the variability of the values for each author. As an example, suppose we found the results for mean sentence lengths given in Table 3.2.

For the example data, the disputed session's mean sentence length is 13, which is quite unusual for Author 2, whose values are consistently between 9 and 11. A value of 13 is also somewhat unusual for Author 1, with values a little more variable, between 8 and 12. A value of 13 is not unusual at all for Author 3, whose values are highly variable, from 6 to 14. So, based on this information, we would lean away from Author 2 and would probably bet most heavily on Author 3. Notice that the mean equals 10 for each of the three authors. The median is also 10. If we make a decision using this data, it is based more on variability than on means.

What if the results seem ambiguous or inconclusive? Suppose one discriminator seems to point to Author 1, but another discriminator points to Author 2. Or, suppose all three authors have apparently similar patterns in mean sentence lengths and similar rates of occurrence for each of your team's chosen discriminators. In that case, no author will stand out clearly as Author 0 based on these discriminators and the limited amount of data we have been able to collect. Still another possibility is that the results may narrow the candidate pool from three to two authors, which is a partially conclusive result. In instances like these, we must be honest with ourselves and with our audience, stating in the report that the results are inconclusive or are only partially conclusive. This is true of any research activity: Often an experiment does not conclusively affirm or disprove all of its motivating hypotheses. Nevertheless, some information is gained and added to the existing beliefs and body of knowledge on the subject.

STEP 5: POLYGON PLOT

A full, formal classification analysis would use not only the patterns in individual discriminator rates, but also the joint pattern in several variables at a time. We will look at the patterns for two variables at a time using a scatter plot. A scatter plot of your team's second discriminator rates versus the first discriminator rates locates a point on the Cartesian plane for each pair of values, with the point's coordinates given by (horizontal coordinate, vertical coordinate) = (Rate1 value, Rate2 value). Note that when we say "Rate2 versus Rate1," we mean Rate2 is to be plotted on the vertical axis.

In this case, our points will be plotted using letters *A, B, C,* and *Z* as plotting symbols. These will identify the authors as *A* = Author 1, *B* = Author 2, *C* = Author 3, and *Z* = Author 0. (Unfortunately, Minitab will not plot the numbers 0, 1, 2, and 3.) We will use a straightedge to carefully draw a polygon around the points having the same letter. The polygon will be a series of connected line segments and will look like a taut rubber band stretched around the points for a particular author. Figure 3.6 illustrates author polygon plots for some example data.

The points in Figure 3.6 labeled *A* correspond to (Rate1, Rate2) pairs for Author 1. The connected *A* points form a polygon that identifies a region of typical (Rate1,

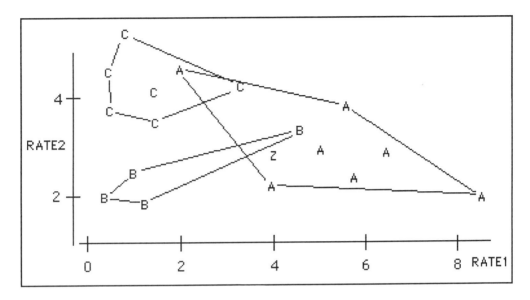

Figure 3.6 Author Polygon Plots for Example Data

Rate2) pairs for Author 1. The *B* polygon identifies Author 2's territory, and the *C* polygon, Author 3's. The point labeled *Z* corresponds to the (Rate1, Rate2) pair for the disputed session. Notice that the *Z* point lies in the middle of the *A* polygon, near the border of the *B* polygon, and well away from the *C* polygon. This polygon plot thus points away from Author 3 and somewhat away from Author 2 as possible authors for the disputed session. Of course, it is certainly possible that the *Z* point lies outside the polygon formed by the actual author's points.

Choose two discriminators to make a polygon plot of your own. You should probably choose the two discriminators that seemed most informative (gave the best authorship clues) based on the dotplot analysis. The following instructions assume that the best discriminators are Rate2 and Rate1. Modify the instructions, replacing these variable names with those of your own choice.

We begin by making a scatter plot of Rate2 versus Rate1:

1. Under the **Graph** menu, select **Scatter Plot**. A Scatter Plot dialog box similar to Figure 3.7 will appear.
2. Click in the box to the right of **Vertical axis** and then double-click on **Rate2**.
3. Click in the box to the right of **Horizontal axis** and then double-click on **Rate1**.
4. Click in the button to the left of **Use tags in**.

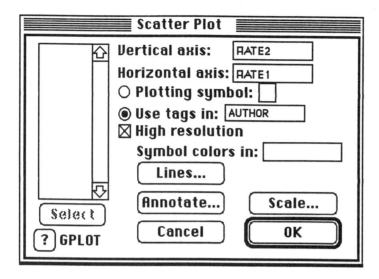

Figure 3.7 Scatter Plot Dialog Box

5. Click in the box to the right of **Use tags in** and then double-click on **Author**.

6. Click **Annotate**. An Annotate Scatter Plot dialog box similar to Figure 3.8 will appear.

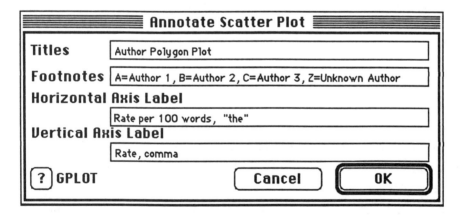

Figure 3.8 Annotate Scatter Plot Dialog Box

7. Click in the first **Titles** box and type **Author Polygon Plot**.

8. Click in the first **Footnotes** box and type *A*=**Author 1,** *B*=**Author 2,** *C*=**Author 3,** *Z*=**Unknown Author**.

9. Click in the **Horizontal Axis Label** box and type **Rate per 100 words,** _____, filling in the blank with the appropriate discriminator word or punctuation for the horizontal axis.

10. Click in the **Vertical Axis Label** box and type **Rate,** _____, filling in the blank with the appropriate discriminator word or punctuation for the vertical axis.

11. Click **OK** to close the Annotate dialog box.

12. Click **OK** to close the Scatter Plot dialog box.

In a moment the Graph window should open and you should see your scatter plot with annotations as suggested. You can resize the plot by resizing the Graph window itself (drag the resize box in the lower-right corner). Print your scatter plot by selecting **Print Window** under the **File** menu.

On your printed graph, use a straightedge to form polygons for the *A* points, the *B* points, and the *C* points. If you have any 2s on the plot, it means two letters are plotted on top of each other at that spot. In this case, you have to go back and look at the rates to find out which letters—*A, B, C,* or *Z*—are there. If there's a 3 on the plot, it means 3 letters are plotted there. Does your author polygon plot shed any light on the question of authorship of the disputed session?

This concludes our calculations on the data. Make sure you have all the output you need, and then quit Minitab by selecting **Quit** from the **File** menu.

If it is convenient, gather together again into your original teams to discuss your findings. Use the "Think About the Data" section (step 4) to aid your reasoning. Who wrote the disputed session: Author 1, Author 2, or Author 3? How sure are you? It is hard to say without formal probability calculations. These would be possible with a sufficiently powerful computer, the right software, and an expert on classification analysis. For our purposes, we can only go with "gut feelings."

PARTING GLANCES

This session is modeled after a famous application of statistics in identifying the authorship of 12 disputed articles in the series known as the *Federalist Papers*. The total series of 85 papers appeared as letters in New York newspapers during 1787 and 1788, published under the pseudonym Publius. The letters attempted to persuade the citizens of the state to ratify the Constitution. It later became known that the papers were collectively written by Alexander Hamilton, John Jay, and James Madison, but for 12 of the 85 papers it was not revealed which of these three was the author. The authorship of these 12 mystery papers was hotly disputed for decades. Applying **discriminant analysis**, a more complicated and formal version of the classification methods discussed in this session, Mosteller and Wallace (1964) found very strong evidence that all 12 of the disputed papers were written by Madison. An excellent and very readable summary of their work can be found in Tanur et al. (1989), a highly recommended book that provides many examples of applying statistics to real-world problems.

There have been many uses of classification analysis in many other fields. Based on a collection of observations, a doctor will classify a patient as diseased or well. In biology, a scientist might classify a skull as belonging to one of several possible species based on measurements of skull length, skull breadth, and so on.

REFERENCES

Mosteller, F., and Wallace, D. L. (1964), *Inference and Disputed Authorship: The Federalist Papers* (Reading, MA: Addison-Wesley).

Tanur, Judith M., Mosteller, Frederick, Kruskal, William H., Lehmann, Erich L., Link, Richard F., Pieters, Richard S., and Rising, Gerald R., ed. (1989), *Statistics: A Guide to the Unknown,* 3rd ed. (San Francisco: Holden-Day).

EXTENDED WRITING ASSIGNMENT

Refer to Appendix 1, "Technical Report Writing," and Appendix 2, "Technical Report Writing Checklist," for guidance on format and style for your report.

Write a report on your team's effort to identify the authorship of the disputed session. Your report should include:

1. A clear and complete description of the problem
2. Rationale for your team's choices of discriminators and operational definitions used in counting
3. Well-labeled plots
4. Conclusions, if any, and rationale for them

Name   Section   Session 3

SHORT ANSWER WRITING ASSIGNMENT

All answers should be complete sentences. Include your dotplots and author polygon plot with this assignment.

1. The data analysis performed in this session is a simplified form of a more formal data analysis designed to help place an object into one of several possible populations. What is this formal type of analysis called?

2. Give a real-world example other than those provided in this session where classification analysis might be useful.

3. What were your team's criteria for choosing discriminators? What were the operational definitions your team adopted to help count sentences and your team's discriminators?

4. Why did we convert the counts to rates per 100 words before analyzing the data?

5. What were your team's conclusions, if any, about the authorship of the disputed session? Discuss these in detail, referring to the plots.

SESSION FOUR

Secrets Behind a Green Thumb

INTRODUCTION

A well-designed experiment is one in which certain variables are manipulated to determine their effects on another variable. Such experiments can be found in all disciplines. In education, designed experiments are used to determine whether a new teaching method is more effective than the standard one. A chemist might plan an experiment to determine whether temperature or amount of catalyst has a significant effect on chemical process readings. In the business world, a market researcher uses experimentation to learn about the effect of various advertising campaigns on the sales of a particular product.

STATISTICAL CONCEPTS

Planned experiments, two-factor design, factor-level selection, randomization.

MATERIALS NEEDED

For each team, 12 small flower pots with identification tags, potting soil, seeds for two varieties of the same vegetable, meter stick, planting trowel, 1/4-cup measure, and a 5-quart bucket of water.

THE SETTING

You are a member of a small agricultural experimentation team working for Harvest Veg, a company that grows vegetables organically without the use of any pesticides. The company produces and distributes several vegetables such as onions, carrots, peas, and tomatoes. Your objective is to design and carry out a small experiment to study the effect of seed type and amount of water on the growth of a particular type of plant. Due to budget, time, and space constraints, you will have only 12 small planting plots and 2 seed varieties. The experiment should last about two months.

BACKGROUND

In an effort to produce large and healthy vegetables, Harvest Veg normally grows their plants in dark, rich soil. Recently, some field technicians have noticed considerable variability in the size of the plants. Some plants appear healthy and large, while others grown in the same field are very small.

There could be several explanations for the varying plant sizes. Seed variety is one. Your research team oversees several small fields that contain various varieties of the same vegetable. For example, half the field may contain one variety of peas and the other half a second variety. It could simply be the case that one variety grows better than the other.

A state-of-the-art irrigation and sprinkler system was recently put in place. However, there seems to be no set system or schedule for watering the Harvest Veg fields. The sprinklers are simply turned on when the plants "look dry." This haphazard method of watering may be another explanation for the varying plant sizes.

You wish to determine the reason for the varying plant sizes. It would be extremely advantageous to Harvest Veg if the occurrence of smaller, less healthy plants was eliminated altogether.

THE EXPERIMENT

STEP 1: SELECTION OF FACTORS

One could use several variables to describe a plant's growth. Weight, height, and number of leaves are just a few. In this experiment, we use plant height, measured each week. In the lingo of designed experiments, we say that the **response variable** or **dependent variable** is plant height.

To simulate Harvest Veg's rich soil, we shall use a top-grade potting soil. Each research team will have two controllable variables: seed variety and amount of water. These variables are called the **factors** or the **independent variables**.

Table 4.1 Levels of Seed and Water Factors

Factor	Level	Description
Seed	1	
	2	
Water	1	
	2	

The seed factor will be two varieties of the same type of plant, and its levels will be denoted by 1 and 2. Write the names of these two varieties in Table 4.1. Your team should decide on two reasonable watering plans. Keep in mind that these two watering plans will be used throughout the experiment. An example of possible water-factor levels is:

Level 1 of water is 1/3 cup of water twice a week.

Level 2 of water is 1/2 cup of water twice a week.

Another example is:

Level 1 of water is 1/2 cup once a week.

Level 2 of water is 1/3 cup twice a week.

Write your two levels of the water factor in Table 4.1.

The four treatment combinations will be denoted by A, B, C, and D, as indicated in Table 4.2. Complete the Detailed Description column of Table 4.2 using the levels of the seed and water factors that you specified in Table 4.1. We wish to determine which of these four combinations produces the largest plants. As 12 pots are available, we shall have 3 pots for each treatment combination. This is an example of a **two-factor experiment with replication**.

STEP 2: RANDOMIZATION

Although an effort has been made to control for soil type, water, and seed variety, there are invariably other factors affecting plant growth that cannot be controlled. For example, we cannot control the amount of light coming through the window. The ideal

Table 4.2 Treatment Combinations for Plant-Growth Experiment

Symbol	Levels	Detailed Description	General Description
A	(1, 1)		Seed variety 1 and water level 1
B	(1, 2)		Seed variety 1 and water level 2
C	(2, 1)		Seed variety 2 and water level 1
D	(2, 2)		Seed variety 2 and water level 2

situation would be for all 12 plants to receive the same amount of light, so any differences in plant growth will be due to the two controlled factors of water and seed variety. List another **uncontrollable factor** besides light in our plant-growth experiment.

To minimize the effect of uncontrollable factors, it is very important that the levels of the independent variables are assigned at random to the experimental units, the pots, in the study. **Randomization** is a technique for assigning treatment combinations to experimental units (in this case, pots). If your 12 pots were set up on a windowsill in one long row, the arrangement pictured in Figure 4.1 does not appear to be random. What if the left-hand side of the window received the most light? Do you see why the assignment in Figure 4.1 would not be a good one?

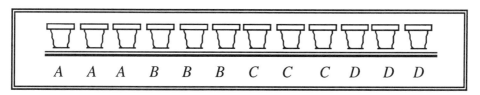

Figure 4.1 Example of a Nonrandomized Experimental Design

Randomization is carried out using random numbers. We will use Minitab to generate a column of random numbers to assign the treatment combinations to each pot. These random numbers have the property that each of the values 1, 2, 3, or 4 is equally likely to occur.

Launch Minitab. To generate 30 random integers between 1 and 4 using Minitab:

1. Under the **Calc** menu, click and hold on **Random Data** and then select **Integer** from the submenu. An **Integer Distribution** dialog box similar to Figure 4.2 will appear.

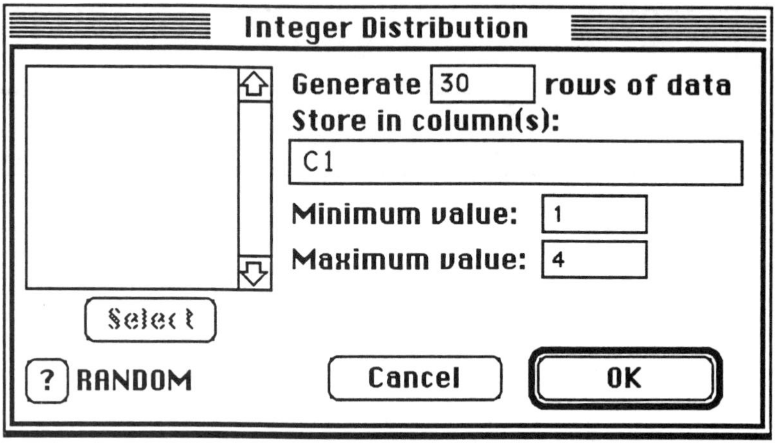

Figure 4.2 Integer Distribution Dialog Box

2. Click in the box to the right of **Generate** and type **30**.
3. Click in the box beneath **Store in column(s)** and type **C1**.
4. Click in the box next to **Minimum value** and type **1**.
5. Click in the box next to **Maximum value** and type **4**.
6. Click **OK**.//

After a short period of time the random numbers will appear in column C1 of the worksheet. The numbers correspond to the treatment combinations as shown in Table 4.3.

The first number in C1 will assign a treatment combination to the first pot. Continue using the numbers in C1 until all 12 pots have been assigned a treatment combination. Remember, each treatment is to be used only three times. That is, after three 1s have appeared, skip over the remaining 1s. Use Figure 4.3 to record the

Table 4.3 Random Number Assignment Rule

Random Number	Treatment Combination
1	A (1, 1)
2	B (1, 2)
3	C (2, 1)
4	D (2, 2)

treatment-combination assignments as they are determined. When you are finished, you should have 12 pots labeled as 3 *A*s, 3 *B*s, 3 *C*s, and 3 *D*s. You probably will not need all 30 generated random integers. Also record the treatment-combination assignments at the top of Table 4.4. After you have completed the assignment of treatment combinations, quit Minitab by selecting **Quit** under the **File** menu.

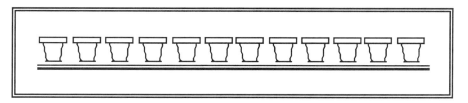

Figure 4.3 Treatment-Combination Assignments to Pots

Don't be alarmed if the numbers in column 1 do not "look" random. Because you used Minitab, you can be assured that they are. It just so happens that with only four different values (1 through 4), it may be that three 4s appear together. This does not mean the list of numbers is not random. If we had generated random integers with values between 1 and 100, observing three 4s together would be very unlikely, and we might doubt the randomness of such data.

STEP 3: PLANTING THE SEEDS

We are now ready to plant the seeds. For each pot, make sure the proper treatment combination is being used. Label each pot with an identification tag that indicates the seed type and watering plan. Also, to minimize the effect of the uncontrollable factors,

Table 4.4 Weekly Plant Heights Measured to Nearest 1/4 Centimeter

Week	Pot (indicate treatment combination)											
	1	2	3	4	5	6	7	8	9	10	11	12
1												
2												
3												
4												
5												
6												
7												
8												
9												
10												

make the amount of soil and the position of the seed with regard to depth and distance from edge of pot as consistent as possible.

Set the pots in their locations. Each week at the beginning of the lab you should water your plants and measure the height of each plant to the nearest 1/4 centimeter. *Do not deviate from your set watering schemes,* even if the plants do not appear healthy.

The weekly heights should be entered in the data collection sheet (Table 4.4). Do not be careless about measuring height. Think about how height will be measured and stick with that rule. For example, if your plant is of the "droopy" variety, then you might choose to straighten out the plant before the height measurement is taken. If so, you should do this every time the height is measured. We analyze this data in Session 12.

PARTING GLANCES

A designed experiment was begun to investigate differences in plant growth. Another way of stating this is that we are looking for sources of variability in the plant heights. At the end of the experiment, the 12 plants will all have different heights. Studying the variability in these 12 measurements and what contributed to this variability is what designed experimentation is all about.

Not all factors turn out to be significant. It may be that your two levels of the water factor produce plants of very similar size. This is fine and would imply that with your chosen levels, water is not a significant source of variability.

Although one of our four factor-level combinations is likely to 'beat" the other three, it may be that we still haven't discovered the optimal setting. But our experiment might suggest other levels that should be investigated in further experimentation. We address these issues further in Session 12.

EXTENDED WRITING ASSIGNMENT

Refer to Appendix 1, "Technical Report Writing," and Appendix 2, "Technical Report Writing Checklist," for guidance on format and style for your report.

Harvest Veg President Irene Bean wants a report summarizing the initial phase of this experiment. Your report should include:

1. Statement of problem and purpose of experiment

2. Selection of response variable and the factors

3. The importance of randomization and how it was accomplished

You should include diagrams and anything else that helps to clarify your description of the plant-growth experiment.

Name _____ Section _____ Session 4

SHORT ANSWER WRITING ASSIGNMENT

All answers should be complete sentences.

1. What is the main goal of this experiment?

2. What were the two seed varieties for your experiment?

3. What are the two levels of the water factor? Briefly describe the thought process that went into selecting these levels.

4. What is the response variable? How often will it be measured and in what units?

5. How was randomization used in this experiment?

6. List a factor other than seed variety and water that could be used in this plant-growth experiment. Suggest two levels for this factor.

7. List another response variable that could have been used.

8. Give an example of an uncontrollable factor in the current setup. Do you anticipate that this factor will significantly affect your findings?

SESSION FIVE

Real and Perceived Distances

INTRODUCTION

One of the most important aspects of data analysis is the study of relationships between variables. How does a cricket's chirping rate change as temperature decreases? How does the yield of a chemical reaction change when pressure is increased? This session introduces graphical and descriptive tools to help quantify relationships between variables.

STATISTICAL CONCEPTS

Scatter plot, regression, calibration, bias, measurement error, variability within and between individuals.

MATERIALS NEEDED

For each team, a 50-foot tape measure and a straightedge.

THE SETTING

Often the measurement we really wish to make on an object is difficult to make (or expensive, toxic, or destructive). If we can find an easily measured variable that is closely linked to the difficult one, we may be able to use the easy one in place of the difficult one. Before doing so, we should do an experiment, called a **regression experiment**, that involves measuring both variables on each of several objects and then studying the manner in which the easy measurement tends to change with the difficult one. We may then be able to adjust the easy measurement to better approximate the difficult one. This process is called **calibration**. This session is a regression and calibration experiment to study the manner in which guessed distances between objects (an easy measurement) change in response to true distances (a more difficult measurement).

BACKGROUND

It is well known that people tend to underestimate the size of faraway objects. Do we also tend to underestimate the distance to faraway objects, or do we tend to overestimate these distances? Or, do we guess right, on average?

THE EXPERIMENT

STEP 1: DATA COLLECTION

The class as a group will go to a pre-chosen spot, with lab manuals and pencils. The instructor will identify a fixed reference point, such as a street sign. He or she will then identify a landmark. You should write a brief description of this landmark at the top of column 2 of Table 5.1. The class will then be asked to guess the distance between the reference point and the landmark. *Please keep your guess to yourself* so as not to influence others. To simplify calculations later, guess in units of feet only. Silently record your guessed distance in column 3 of Table 5.1. Then the instructor will ask you to guess the distance between the reference point and a second landmark, to be recorded in Table 5.1, and then another, and so on, for a total of 13 landmarks. Don't worry that your guesses might be bad. That's variability, and it's what we are studying.

The class will then be split into teams to measure the *true* distances to the landmarks. Each team will have three members:

1. The Base: This person holds the tape end at the reference point, and at intermediate points along the way if the landmark is too far away to measure in one tape length. He or she also advises the other team members if they are not walking straight toward the landmark and keeps track of the number of full tape lengths that have been used en route to the landmark.

Table 5.1 Distances Between a Fixed Point and Several Landmarks

Landmark Number	Landmark Description	Guessed Distance (feet)	Measured Distance (feet)	Median Measured Distance
1				
2				
3				
4				
5				
6				
7				
8				
9				
10				
11				
12				
13				

2. The Point: This person takes the tape roll and carefully walks straight toward the landmark, until it is reached or the tape runs out. If the tape runs out, the Point is responsible for keeping track of exactly where the starting point for the next tape length will be while the Base comes forward. Also, the Point verifies the final reading that the Eyes makes.

3. The Eyes: This person walks beside the Point. When the landmark is reached, he or she reads the tape and (in conjunction with the Base and the Point) calculates the final measured distance to the landmark and records it in the appropriate row of Table 5.1, column 4. The Eyes is also the spokesperson for the team in class discussion.

Each of the first 12 landmark distances will be independently measured by at least 3 teams. There are two serious errors that occur with surprising frequency:

1. It is very easy to forget how many tape lengths have been used when measuring distant landmarks. It is the Base's responsibility to remember this, but the other team members should help, too.

2. If the end of the tape is reached, the Point should be very careful where the start of the next tape length is marked. For example, if you are using 50-foot tape measures, the tape is actually longer than 50 feet, but *the new start point should be at the 50-foot mark, not the tape end.* The Eyes should back up the Point to help prevent this error.

Do not measure the 13th landmark distance. It will be a test case. Its true distance has been measured in advance by your instructor, and we discuss it later.

STEP 2: INDIVIDUAL DATA ANALYSIS

First, the instructor will lead the class in resolving team-to-team differences in measured distances. The median of all the measured distances for each landmark will be used as the *true* distance. Fill in column 5 of Table 5.1 with these medians as the discussion proceeds. Notice that, by using the median of at least three measurements, if one of the teams messed up in a big way, its mistake will not have much effect on the final number.

Turn on your Mac and launch Minitab. In the new worksheet, give columns C1, C2, and C3 the variable names **Landmark**, **Guess**, and **True**. Figure 5.1 illustrates the variable names in the Untitled worksheet. Carefully enter your data for the first 12 landmarks from Table 5.1 (columns 1, 3, and 5) for these variables. *Do not enter data for the 13th landmark.* After checking your data carefully, save the worksheet onto your diskette by selecting **Save Worksheet As** from the **File** menu. Name the worksheet **distances(mine)**.

We are now ready to examine the relationship between the true distances and your guessed distances. The most useful graphical tool for examining the relationship between two variables is the **scatter plot**. A scatter plot of guessed distances versus true distances locates a point on the Cartesian plane for each landmark, with the point's coordinates given by (horizontal coordinate, vertical coordinate) = (true distance, guessed distance). Note that when we say "guessed distance versus true distance" we mean that guessed distances are to be on the vertical axis. We can

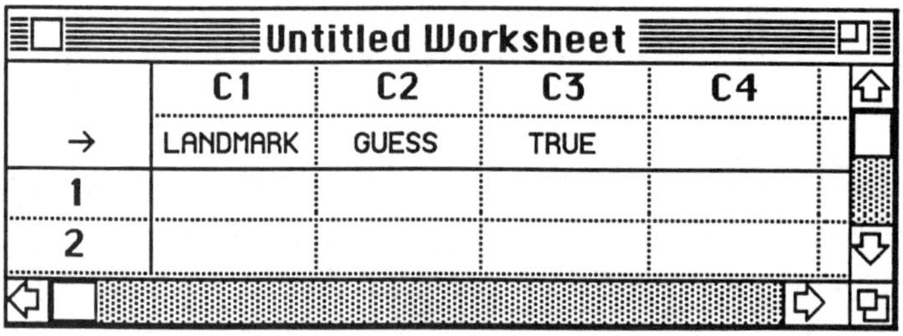

	C1	C2	C3	C4
→	LANDMARK	GUESS	TRUE	
1				
2				

Figure 5.1 Variable Names in Untitled Worksheet

make a scatter plot, with an added 45° line to help you judge whether you are an accurate guesser, as follows:

1. Under the **Graph** menu, select **Scatter Plot**. A Scatter Plot dialog box similar to Figure 5.2 will appear.

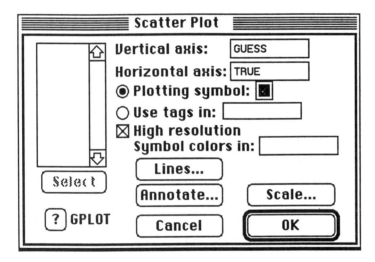

Figure 5.2 Scatter Plot Dialog Box

2. Click in the box to the right of **Vertical axis** and type **Guess**.

3. Click in the box to the right of **Horizontal axis** and type **True**.

4. Click **Annotate**. An Annotate Scatter Plot dialog box similar to Figure 5.3 will appear.

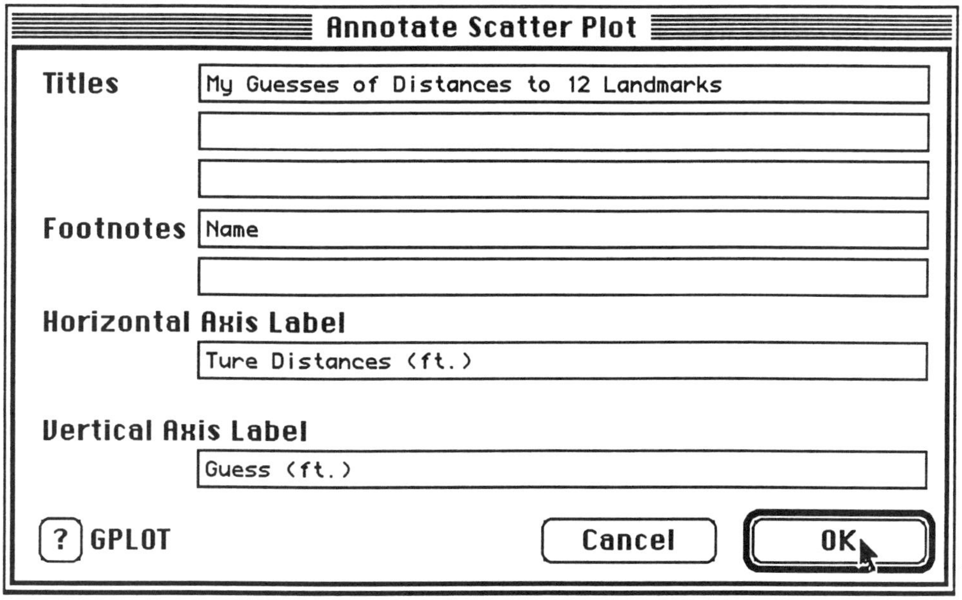

```
╔════════════ Annotate Scatter Plot ════════════╗
║                                                ║
║  Titles    [My Guesses of Distances to 12 Landmarks]
║            [                                  ]
║            [                                  ]
║                                                ║
║  Footnotes [Name                              ]
║            [                                  ]
║                                                ║
║  Horizontal Axis Label                         ║
║            [Ture Distances (ft.)              ]
║                                                ║
║  Vertical Axis Label                           ║
║            [Guess (ft.)                       ]
║                                                ║
║  [?] GPLOT              [ Cancel ]  [  OK  ]  ║
╚════════════════════════════════════════════════╝
```

Figure 5.3 Annotate Scatter Plot Dialog Box

5. Click in the first **Title** box, and type **My Guesses of Distances to 12 Landmarks**.

6. Click in the first **Footnotes** box and type your name.

7. Click in the **Horizontal Axis Label** box and type **True Distance (ft.)**.

8. Click in the **Vertical Axis Label** box and type **Guess (ft.)**.

9. Click **OK** to end the Annotate command.

10. To add a 45° line to the plot, click **Lines**. A Lines dialog box similar to Figure 5.4 will appear.

11. Click in the first box under **Y column** and type **True**.

12. Click in the first box under **X column** and type **True** again.

13. Click **OK** to end the Lines command.

14. Click **OK** to end the Scatter Plot command.

The choices in the Lines dialog box are admittedly mysterious, but they tell Minitab to draw a 45° line on the plot. This is because the 45° line is defined by points (x, y)

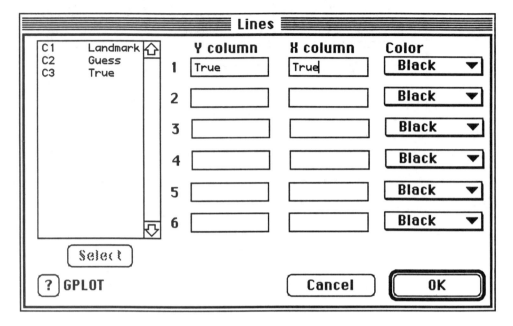

Figure 5.4 Lines Dialog Box

having $y = x$, so we can plot the line simply by using the same variable (any variable) for both X and Y in the Lines command. In this case we used the variable True, but any numeric variable could have been used.

After a short delay, the Gplot 'Guess' 'True' window should open, and you should see your scatter plot with annotations as suggested and the 45° line, analogous to Figure 5.5. Print your scatter plot at this time by selecting **Print Window** under the **File** menu. After you get your printout, quit Minitab by selecting **Quit** under the **File** menu.

STEP 3: CALIBRATION

Do the points on your own scatter plot lie approximately on a straight line? If so, the relationship between your guessed distances and the true distances is said to be approximately **linear**. Do the points seem to describe a curve? If there seems to be a U-shaped curve, the relationship is said to be **convex**. If it is an inverted U-shape, it is said to be **concave**. How would you describe the relationship shown by the points on your plot?

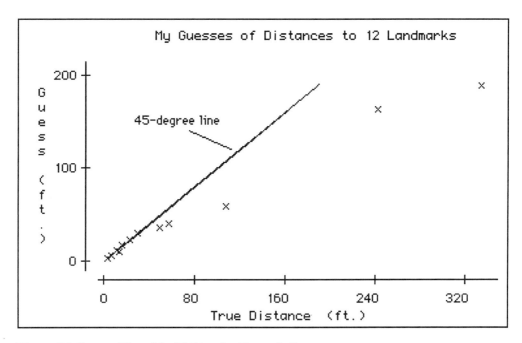

Figure 5.5 Scatter Plot with 45° Line for Example Data

Ignoring the 45° line, draw the best straight line, smooth convex curve, or smooth concave curve that you can through the center of the point cloud in your scatter plot. Use the straightedge if you decide to draw a straight line. If you draw a curve, make it a smooth curve through the center of the point cloud—don't just connect the dots. What you have just done is sketched an approximate regression line or curve, a line or curve that helps to summarize the relationship between the two variables.

If this sketched line or curve lies below the 45° line, your guesses tend to underestimate the true distances to landmarks. We would then say that as a guesser you are **negatively biased**. This is the case for the majority of guessers, including the one whose data is shown in Figure 5.5. Some guessers are fairly accurate on the average with their guesses. That is, the points on their scatter plot tend to fall along the 45° line. We say they are **unbiased** guessers. A rare few individuals tend to overestimate; the points on their scatter plot tend to lie above the 45° line. We say they are **positively biased** guessers. Of course, one's ability as a guesser may vary from situation to situation.

If we consider you, as a distance guesser, to be a new sort of measuring instrument, we can use the sketched regression line or curve on your plot to **calibrate** you. Calibration is an activity or operation for correcting bias in a measuring device and is

88   SESSION FIVE

an example of one very important use for regression experiments. We will say you are calibrated as a distance guesser if you always adjust your initial guess in the following way:

1. Locate the point on the vertical axis of your plot that corresponds to your initial guess.

2. With a straightedge, draw a horizontal line from that point across the plot until the line touches the sketched regression line or curve.

3. From that point, drop a vertical line to the horizontal axis.

4. The adjusted guess is the reading on the horizontal axis found at the end of this vertical line.

Figure 5.6 shows the result of using the calibration to adjust an initial guess of 100 feet, using the example data from Figure 5.5. The calibration leads to an adjusted guessed distance of 160 feet in this case, an increase of 60 feet from the initial guess. This makes a lot of sense under the observation that this guesser was negatively biased; if the initial guess is 100 feet, it should be adjusted upward.

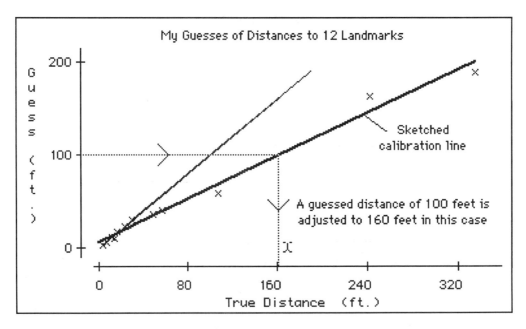

Figure 5.6 Calibration with Example Data

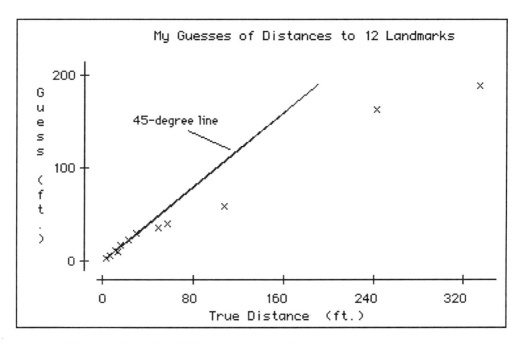

Figure 5.5 Scatter Plot with 45° Line for Example Data

Ignoring the 45° line, draw the best straight line, smooth convex curve, or smooth concave curve that you can through the center of the point cloud in your scatter plot. Use the straightedge if you decide to draw a straight line. If you draw a curve, make it a smooth curve through the center of the point cloud—don't just connect the dots. What you have just done is sketched an approximate regression line or curve, a line or curve that helps to summarize the relationship between the two variables.

If this sketched line or curve lies below the 45° line, your guesses tend to underestimate the true distances to landmarks. We would then say that as a guesser you are **negatively biased**. This is the case for the majority of guessers, including the one whose data is shown in Figure 5.5. Some guessers are fairly accurate on the average with their guesses. That is, the points on their scatter plot tend to fall along the 45° line. We say they are **unbiased** guessers. A rare few individuals tend to overestimate; the points on their scatter plot tend to lie above the 45° line. We say they are **positively biased** guessers. Of course, one's ability as a guesser may vary from situation to situation.

If we consider you, as a distance guesser, to be a new sort of measuring instrument, we can use the sketched regression line or curve on your plot to **calibrate** you. Calibration is an activity or operation for correcting bias in a measuring device and is an example of one very important use for regression experiments. We will say you are calibrated as a distance guesser if you always adjust your initial guess in the following way:

1. Locate the point on the vertical axis of your plot that corresponds to your initial guess.
2. With a straightedge, draw a horizontal line from that point across the plot until the line touches the sketched regression line or curve.
3. From that point, drop a vertical line to the horizontal axis.
4. The adjusted guess is the reading on the horizontal axis found at the end of this vertical line.

Figure 5.6 shows the result of using the calibration to adjust an initial guess of 100 feet, using the example data from Figure 5.5. The calibration leads to an adjusted guessed distance of 160 feet in this case, an increase of 60 feet from the initial guess. This makes a lot of sense under the observation that this guesser was negatively biased; if the initial guess is 100 feet, it should be adjusted upward.

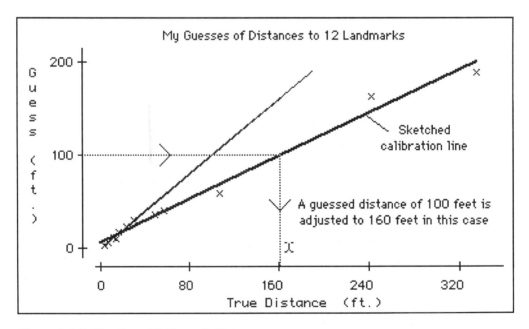

Figure 5.6 Calibration with Example Data

Now, you try it. For the mystery 13th landmark, the one whose true distance was not measured, use your guessed distance and your sketched regression line or curve to arrive at an adjusted guessed distance. Your instructor will tell you the true distance. Did the calibration adjustment improve your guess?

STEP 4: GROUP DATA ANALYSIS

Write your name on your Table 5.1 and give it to the instructor. While he or she is getting organized, choose a Macintosh, launch Minitab, and name the first three columns of a new worksheet **Landmark**, **Guess**, and **True** as you did in Step 2 for your own data. Save the empty worksheet on a diskette as you did in Step 2, except name it **Distances(class)**.

The instructor will then read data to you in groups of three values. The values read will be the landmark number, guessed distance, and true distance for several randomly chosen guessers for each landmark. Type these carefully in rows in your data set as the instructor reads them to you. When all the data has been entered and double-checked, you must again save the complete data set by pushing **<Command>+S**.

Make a scatter plot of guessed distances versus true distances for the class's data as you did in Step 2, but modify the title to be **Class's Guesses of Distances to 12 Landmarks**. You'll be asked to comment on this graph in the writing assignment for this session.

Print your scatter plot by selecting **Print Window** under the **File** menu. Then quit Minitab by selecting **Quit** under the **File** menu and, if necessary, copy the data set **Distances(class)** to each computer partner's diskette, as described near the end of Session 1.

PARTING GLANCES

In this experiment, we used an informal regression line or curve to calibrate a "measuring instrument," in this case, a human being guessing distances between objects. A more important calibration exercise was performed to improve verification of nuclear weapons tests under the Threshold Test Ban Treaty between the United States and Russia (Picard and Bryson, 1992). After the cold war, the two countries embarked on an effort to make onsite yield measurements of each other's nuclear tests, for the purpose of calibrating a monitoring system based on seismic measurements. That is, there are two measurements of a nuclear explosion's force:

1. The onsite measurement

2. The seismic disturbance as measured by a seismograph halfway around the world

Once a reliable calibration method is constructed, each country should be able to monitor the other's nuclear tests using at-home seismic measurements, instead of traveling overseas to make onsite measurements. Note that this difficult onsite measurement will become impossible if relations between the United States and Russia return to cold war levels. Also, as more is learned about the relationships between the two measuring methods, nuclear testing in countries other than Russia may be more reliably monitored in the United States by seismologists.

REFERENCE

Picard, Richard, and Bryson, Maurice (1992), "Calibrated Seismic Verification of the Threshold Test Ban Treaty," *Journal of the American Statistical Association* 87, 293–299.

EXTENDED WRITING ASSIGNMENT

Refer to Appendix 1, "Technical Report Writing," and Appendix 2, "Technical Report Writing Checklist," for guidance on format and style for your report.

Write a report describing the measuring and calibration experiment. The report should include:

1. A clear description of the experiment, including how the data was collected

2. A discussion of the pattern in your own guesses, including at least one calibration calculation

3. A discussion of the patterns you see in the guesses of the class as a whole

Use the terms provided under the "Statistical Concepts" section wherever they are appropriate. Incorporate your scatter plots into your report. Use answers to the short-answer writing assignment to help guide your discussion.

Name _____ Section _____ Session 5

SHORT ANSWER WRITING ASSIGNMENT

All answers should be complete sentences. Include scatter plots of your data and the class's data with this assignment.

1. What factors might cause your ability as a guesser to vary from situation to situation?

2. Do the points on your own scatter plot lie approximately on a straight line, a convex curve, a concave curve, or in some other pattern?

3. Do your own distance guesses tend to be negatively biased, positively biased, or approximately unbiased?

4. What was your initial guess for the mystery 13th landmark? What was your adjusted guess after calibration? Did the adjustment lead to a more accurate guess in this case?

5. Is the relationship between guessed distances and true distances for the class as a whole approximately linear, convex, concave, or something else?

6. If we obtained guessed distances by randomly picking a class member and then asking him or her to guess a distance, the class's scatter plot is appropriate for calibrating this measurement. Is this measurement negatively biased, positively biased, or something else?

7. What notable differences do you see between the scatter plot of your guesses and the plot for the group as a whole?

SESSION SIX

Collecting Data over Time

INTRODUCTION

Observations taken sequentially over time are called **time series**. Usually the variable of interest is numerical and is measured at equally spaced units of time, such as monthly, hourly, or yearly. The data in Figure 6.1 represents the number of international airline passengers (in thousands) recorded monthly during the years 1954 to 1960 as reported by Box and Jenkins (1976). Note that the plot clearly shows that airline travel is highest during June and July, which on the graph are the points where Month = 7, 19, 31, etc.

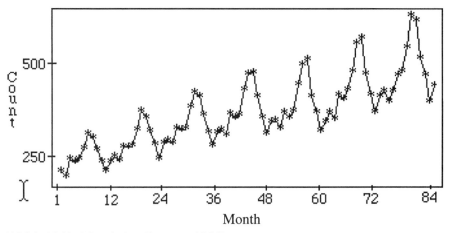

(1954–1960, Month 1 = January 1954)

Figure 6.1 Airplane Passenger Counts, in Thousands (Box and Jenkins [1976])

Once a time series has been collected, it is often of interest to predict future observations. To obtain good predictions, one first investigates whether the data contain certain features, such as seasonal variations, trends, and cycles. **Seasonal variations** are regular changes that recur throughout the data. Daily temperature readings are an example that exhibits strong seasonal behavior, with higher temperatures in the summer and lower ones during the winter. **Trends** are persistent, long-term rises or falls in the data. Note that the airline data of Figure 6.1 show a strong seasonal component with a general upward trend, indicating that airline travel increased from 1954 to 1960. Up and down movements of somewhat regular duration and strength are called **cycles**. Hourly oceanic temperature readings can appear somewhat cyclical because they are related to the tidal cycle.

In this session, you will collect data for a particular time series and learn about some graphical and descriptive tools that are useful in the statistical analysis of time series data.

STATISTICAL CONCEPTS

Time series, trends, cycles, seasonal variation, smoothing a time series.

MATERIALS NEEDED

A stopwatch for each team.

THE SETTING

You are a member of a team that has been hired by the Department of Highways and Transportation to study traffic flow near your school. Your goal is to collect time series related to the traffic flow at different points around campus. You will then investigate the data for trends, cycles, and/or seasonal-type variation and present your findings and recommendations to the highway department.

BACKGROUND

Preliminary studies suggest that traffic flow for some streets on or near the campus is much higher than expected. Traffic at many heavily traveled intersections is partially controlled by stoplights, but it appears that the timing of the lights may need adjusting.

THE EXPERIMENT

STEP 1: CHOOSE A LOCATION

Choose a location where there is a traffic light at one end of the block. You should stand at the middle of the block and observe traffic flow in only one direction—preferably cars coming from the location of the stoplight. You will be counting cars every 15 seconds, so you need to choose a well-traveled road.

It is best to stand at a point where there are no nearby side streets that may interfere with the traffic counts. In Figure 6.2, if you are observing vehicles traveling east, then location A is better than locations B or C. Do you see why?

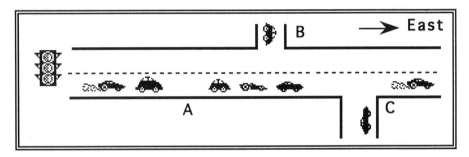

Figure 6.2 Example Locations

Briefly describe your team's location:

STEP 2: COLLECTION OF DATA

Go to your chosen location. The data you should collect is the total number of motor vehicles that pass by in consecutive 15-second time intervals. Count automobiles, motorcycles, and trucks (not bicycles or mopeds) that are coming in only one direction. As one count ends, you should immediately begin counting for the next time period; that is, do not let any motor vehicles go uncounted.

Table 6.1 Practice Counts

Time period 1: _____ Time 2: _____ Time 3: _____ Time 4: _____

The data collection may take some practice, and it may be helpful to assign team members specific duties. One person may run the watch, calling out a cue every 15 seconds. The other person(s) can count and record the data. Get a minute's practice by taking four test observations and recording them in Table 6.1.

After you have mastered the method, you should begin formal collection of the data. You should record 60 counts. It is important that no breaks occur in the data collection process. That is, you should not miss an interval. The statistical procedures that will be used later do not allow for missing values in time series data. Pretending that the missing value did not occur and just writing the next 15-second count in its place is incorrect and can severely distort your time series. After the first minute, your data might look something like this:

Time 1 (first 15-second interval): 11 cars

Time 2 (second 15-second interval): 9 cars

Time 3 (third 15-second interval): 3 cars

Time 4 (fourth 15-second interval): 1 car

Remember, each "time" is a 15-second interval, so the entire data collection process should take about 15 minutes. The time periods are denoted by $t = 1, 2, ..., 59, 60$. Record the observations in Table 6.2.

Before you go inside, time the traffic light at the intersection. How often does it change? What are the durations of the red and green lights? This information may help explain any trends or cycles you see in the data.

STEP 3: DATA ANALYSIS

Insert a diskette. Launch Minitab and enter the data in the first two columns using the variable names **Time** and **Count**. Figure 6.3 shows the Data window for some example data. After the data has been entered and checked, save your worksheet by selecting **Save Worksheet As** under the **File** menu and following the prompts for saving files provided in Session 1. Name the worksheet **Cars**.

Table 6.2 Vehicle Count Data

Time	Count	Time	Count	Time	Count
1		21		41	
2		22		42	
3		23		43	
4		24		44	
5		25		45	
6		26		46	
7		27		47	
8		28		48	
9		29		49	
10		30		50	
11		31		51	
12		32		52	
13		33		53	
14		34		54	
15		35		55	
16		36		56	
17		37		57	
18		38		58	
19		39		59	
20		40		60	

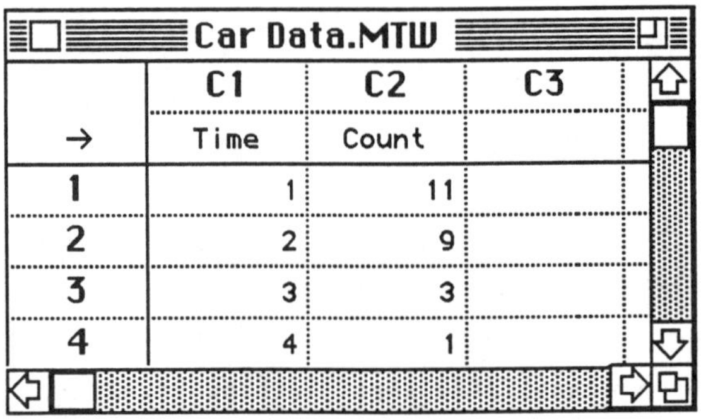

```
╔═══════ Car Data.MTW ═══════╗
║       │   C1   │   C2   │   C3   ║
║   →   │  Time  │  Count │        ║
║   1   │    1   │   11   │        ║
║   2   │    2   │    9   │        ║
║   3   │    3   │    3   │        ║
║   4   │    4   │    1   │        ║
╚═══════════════════════════════════╝
```

Figure 6.3 Data Window with Example Data

We will begin the data analysis by plotting the time series.

1. Under the **Graph** menu, select **Scatter Plot**. A Scatter Plot dialog box similar to Figure 6.4 will appear.

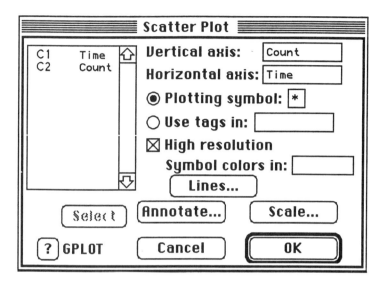

Figure 6.4 Scatter Plot Dialog Box

2. Click in the box next to **Vertical axis** and type **Count**.

3. Click in the box next to **Horizontal axis** and type **Time**.

4. Highlight the **x** in the box next to **Plotting symbol** and type *. You can use any character for the plotting symbol such as a period or an alphabetic character, but the asterisk works well.

5. To connect the data points with lines, click the **Lines** button. A Lines dialog box similar to Figure 6.5 will appear.

6. Click in the first box under **Y column** and type **Count**.

7. Click in the first box under **X column** and type **Time**.

8. Click **OK** to close the Lines dialog box.

9. To add a title and axis labels, click on the **Annotate** button. An Annotate Scatter Plot dialog box similar to Figure 6.6 will appear.

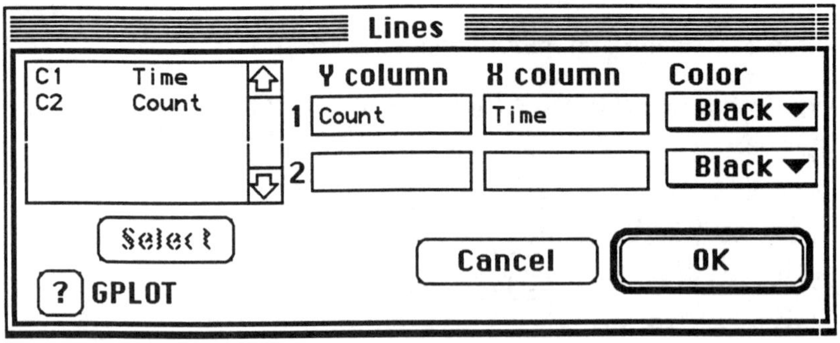

Figure 6.5 Lines Dialog Box

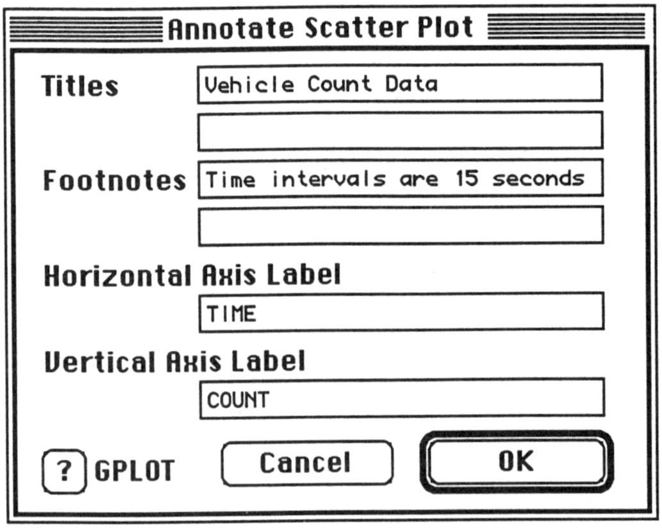

Figure 6.6 Annotate Scatter Plot Dialog Box

10. Click in the first box next to **Titles** and type a descriptive title for the scatter plot. Likewise, click and type to create detailed labels for the horizontal and vertical axes. You may also include a footnote.

11. Click **OK** to close the Annotate dialog box.

12. Click **OK** to produce the scatter plot.

After a short delay, the high-resolution graphics scatter plot will appear. It is important that you print the plot now by selecting **Print Window** under the **File** menu.

An important feature of time series is the amount of variability in the data. Graphs with data points exhibiting high variability are termed "high frequency" or "noisy" data. If the values slowly drift through time without much change, this is considered a "low frequency" time series. Figure 6.7 shows an example of a high frequency data set and also one with little variability. What can you say about the variability in your car counts?

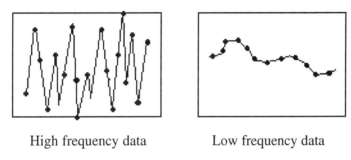

High frequency data   Low frequency data

Figure 6.7 Examples Illustrating Variability in Time Series Plots

Obtaining summary statistics for the time series is simple:

1. Under the **Stat** menu, click and hold on **Basic Statistics** and then select **Descriptive Statistics** from the submenu. A Descriptive Statistics dialog box similar to Figure 6.8 will appear.

2. Double-click on the variable name **Count**.

3. Click **OK**.

The statistics will appear in the Session window. What is the average number of cars passing during a 15-second period? Are most of the counts close to this average? Use the results in the Session window to complete Table 6.3.

The last part of this analysis illustrates how a "bumpy" plot can be smoothed, thus making the cycles or trends more visible. There are many methods of smoothing a time series plot. Here we use what is called **moving average smoothing**, or three-point smoothing. For each time period a new (smoothed) value of the time series is

104   SESSION SIX

Table 6.3 Descriptive Statistics for Car Counts

Mean	Median	Standard Deviation

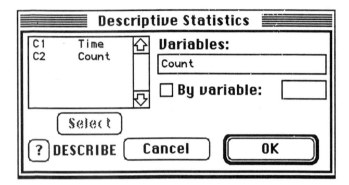

Figure 6.8 Descriptive Statistics Dialog Box

computed, and this new value is simply the average of the data from the previous, current, and next time periods:

$$\text{Smooth } X_t = \frac{X_{t-1} + X_t + X_{t+1}}{3}$$

To get Minitab to compute a smooth value for every time period, we need to define new variables that represent the lagged values of the series. For any observation in the series, the Lag1 value is the previous observation and the Lag2 value is the observation from two periods back.

We will now create the variable Lag1 for the vehicle counts series.

1. Under the **Stat** menu, click and hold on **Time Series** and then select **Lag** from the submenu. A Lag dialog box similar to Figure 6.9 will appear.

2. Click in the box next to **Series** and type **Count**.

3. Click in the box next to **Put lags in** and type **C3**.

4. Click **OK**.

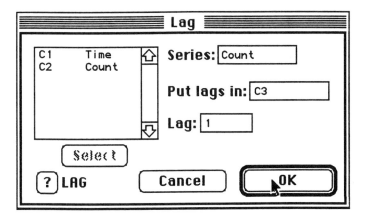

Figure 6.9 First Lag Dialog Box

The Data window should appear with a new column, C3. Click in the first box under **C3** and type the variable name **Lag1**. Look closely at the new variable. There is no value for the first time period, the value of Lag1 at Time 2 equals the value of Count at Time 1, the value of Lag1 at Time 3 equals the value of Count at Time 2, and so on. This is what lagging is all about.

For a three-point moving average, we also need the Lag2 values:

1. Under the **Stat** menu, click and hold on **Time Series** and then select **Lag** from the submenu. A Lag dialog box similar to Figure 6.10 will appear.

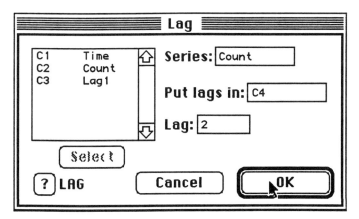

Figure 6.10 Second Lag Dialog Box

2. Click in the box next to **Series** and type **Count**.

3. Click in the box next to **Put lags in** and type **C4**.

4. Highlight the **1** in the box next to **Lag** and type **2**.

5. Click **OK**.

After the Data window appears, name the new variable by typing **Lag2** under **C4**.

Look at the values of **Count**, **Lag1**, and **Lag2** for time 3. What you should see in this row are the vehicle counts for time periods 3, 2, and 1, and so an average of these three values would represent the three-point moving average for time period 2. The mean of the row 4 values represents the three-point moving average of the vehicle counts for time 3, and so on.

Before we get Minitab to compute these row averages, there is something we must first do with the Time variable. Since the average of the row 3 values corresponds to a mean for time period 2, we need a lagged value of the variable Time. Create this variable by repeating the steps for creating Lag1 (see Figure 6.9), using **Time** in place of **Count** and **C5** in place of **C3**. Name the new variable **TLag1**. Figure 6.11 shows the Data screen for the example data after the above manipulations.

	C1	C2	C3	C4	C5
	Time	Count	Lag1	Lag2	TLag1
1	1	13	*	*	*
2	2	9	13	*	1
3	3	2	9	13	2
4	4	1	2	9	3
5	5	0	1	2	4

Figure 6.11 Data Window for Example Data After Manipulation

Finally, before averages for each row can be computed, any rows containing missing observations must be deleted. To do this:

1. Under the **Calc** menu, select **Delete rows**. A Delete Rows dialog box similar to Figure 6.12 will appear.

2. Click in the box next to **Delete rows** and type **1:2**.

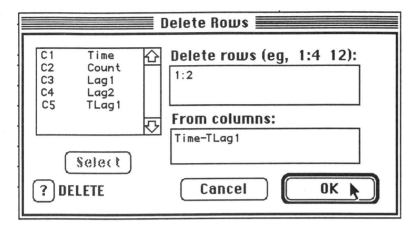

Figure 6.12 Delete Rows Dialog Box

3. Click in the box next to **From columns**. Select all five variables by dragging the pointer across their names until all names are highlighted and then clicking **Select**.

4. Click **OK**.

We are now ready to calculate the three-point moving averages.

1. Under the **Calc** menu, click and hold on **Functions and Statistics** and then select **Row Statistics** from the submenu. A Row Statistics dialog box similar to Figure 6.13 will appear.

2. Click the button next to **Mean**.

3. Select the variables **Count**, **Lag1**, and **Lag2** by double-clicking on each variable name.

4. Click in the box next to **Result in** and type **C6.**

5. Click **OK**.

After a short delay, the moving averages for Count will appear as the new variable in column C6. Name this variable **Smooth**.

A few remarks about smoothing are in order. The variables Lag1 and Lag2 were obtained just to compute the row means and really don't tell us anything additional about the data. (They are the data, almost!) Second, we could have computed medians

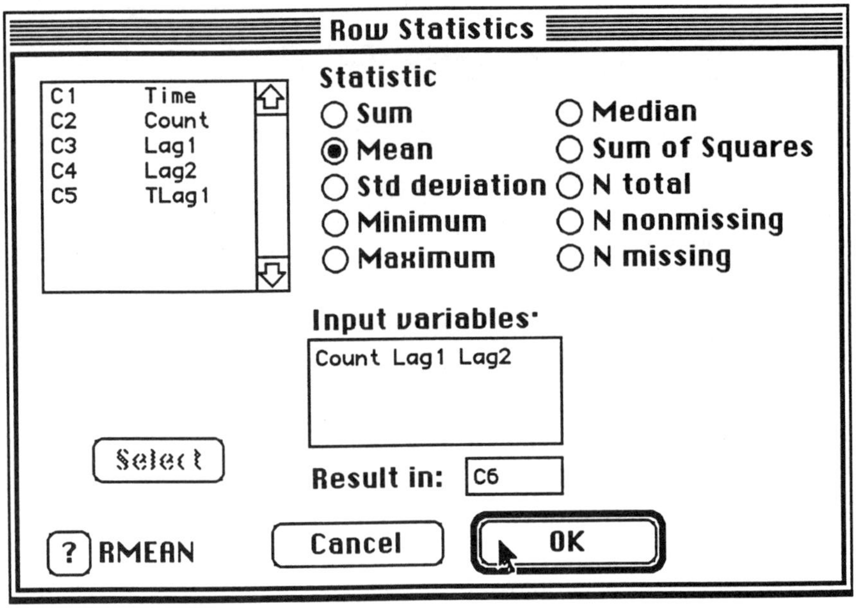

Figure 6.13 Row Statistics Dialog Box

of the three successive data values by simply selecting Median as the statistic of interest under Row Statistics. This would be called the "method of running medians." Finally, we could have computed the average for groups of five observations, but this would have required defining the Lag3 and Lag4 variables for the Count data.

Next construct a scatter plot of the smoothed counts, Smooth, versus the lagged time variable TLag1. The steps are almost identical to those used to obtain our first scatter plot, except you will need to substitute **Smooth** for **Count** and **Tlag1** for **Time**. Remember to use an appropriate title and axis labels. When you are satisfied with the appearance of this plot, print it exactly as you did the first scatter plot.

By smoothing we hope to eliminate some of the "noise" (random, unexplained variation) in the data, thus making it easier to see trends, cycles, or seasonal variation. Did the smoothing eliminate the noise so you can learn more about possible trends and/or cycles in your car counts?

Figure 6.14 shows a smoothed plot of the airline passenger counts. Note that with the elimination of the bumpiness of the original plot, the summer peaks are more

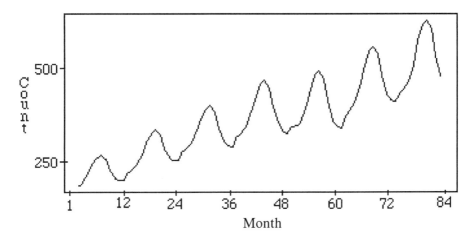

Figure 6.14 Airline Passenger Counts, Smoothed

visible. Smoothing by the moving average method can be thought of as applying a linear filter to the raw data. The smoothing filter is known as a "low pass filter," since it takes the original data and removes the fluctuating, high frequency component and allows the slowly varying, possibly trend or seasonal components in the data to "pass through."

Perhaps it is desirable to create an even smoother plot for your data. Five-point smoothing would be carried out by further defining Lag3 and Lag4 values for the Count variable and then computing row means again. But this approach of creating new lagged variables is a lot of trouble and cumbersome; fortunately, Minitab has some built-in smoothing routines under **EDA** in the Stat menu. EDA, or exploratory data analysis, was made popular by John Tukey (1977) and refers to the use of relatively simple graphical and numerical methods to learn about data. Some of the most widely used EDA methods are the boxplot and the stem-and-leaf plot.

To automatically smooth using Minitab:

1. Under the **Stat** menu, click and hold on **EDA** and then select **Resistant Smooth** from the submenu. A Resistant Smooth dialog box similar to Figure 6.15 will appear.

2. Click in the box next to **Variable** and type **Count**.

3. Click in the box next to **Store rough in** and type **C7**. We will not be concerned with this rough variable, but we must specify a column to hold it.

4. Click in the box next to **Store smooth in** and type **Smooth2**.

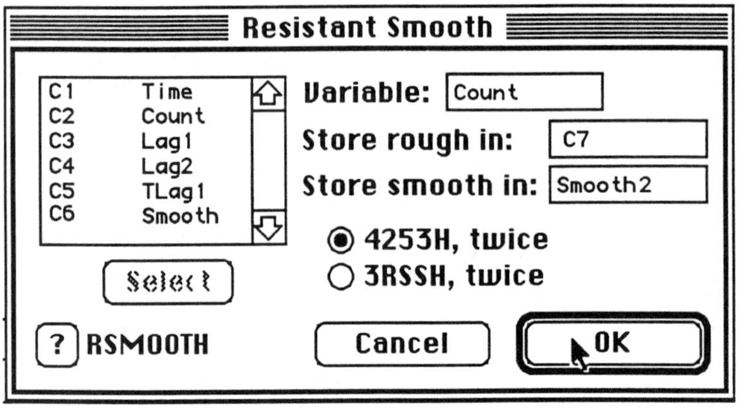

Figure 6.15 Resistant Smooth Dialog Box

5. Click in the box to the left of **4253H, twice**.

6. Click **OK**.

The 4253H smoother is called a resistant smoother because it produces a smooth graph even when the data contains several extreme outliers. It uses running means (you just did a three-point running mean) and running medians. In particular, the smoother first calculates a running median for successive groups of four observations. This is followed by a running median of 2 on this smoothed data. Running medians of order 5 and then 3 are calculated next, and the final step is a Hanning smoothing, or a weighted three-point moving average,

$$\text{Final smooth } X_t = \frac{1}{4} X_{t-1} + \frac{1}{2} X_t + \frac{1}{4} X_{t+1},$$

where the X_t's are the results of the smoothing by running medians 4, 2, 5, and 3. You can see that there is a lot of smoothing going on!

In the last graph we will look at this smoothed data, but we also want to indicate the raw data values. To do this in Minitab:

1. Under the **Graph** menu, select **Multiple Scatter Plot**. A Multiple Scatter Plot dialog box similar to Figure 6.16 will appear.

2. Click in the first box under **Vertical axis** and type **Smooth2**.

3. Click in the first box under **Horizontal axis** and type **Time**.

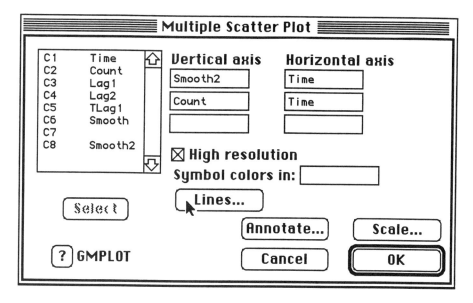

Figure 6.16 Multiple Scatter Plot Dialog Box

4. Click in the second box under **Vertical axis** and type **Count**.

5. Click in the second box under **Horizontal axis** and type **Time**.

6. Click on the **Lines** button. A Lines dialog box similar to Figure 6.17 will appear.

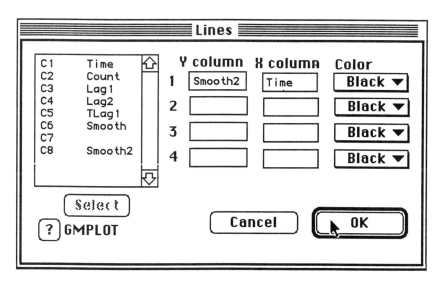

Figure 6.17 Lines Dialog Box

7. Click in the first box under **Y column** and type **Smooth2**.

8. Click in the first box under **X column** and type **Time**.

9. Click **OK** to close the Lines dialog box.

10. Click on **Annotate** and specify appropriate title and axis labels.

11. Click **OK** to close the Annotate dialog box.

12. Click **OK** to produce the plot.

Take a look at this second smoothed data graph. It is possible to oversmooth and therefore lose valuable information present in the data. Which of the two graphs do you think is better? Remember to obtain a printout of your graph by selecting **Print Window** under the **File** menu. This concludes our calculations on the data. Make sure you have all the output you need, and then quit Minitab by selecting **Quit** from the **File** menu.

PARTING GLANCES

In this lab we have collected data for a particular time series and analyzed it using plotting techniques and summary statistics. Further analysis would include more sophisticated techniques such as model fitting to predict future traffic flow.

Time series data occur in many areas. Of interest to a physicist would be the number of particles emitted by a radioactive element during consecutive time intervals, and collecting this data would probably require sophisticated laboratory equipment. In the field of economics, an important variable is the consumer price index (CPI), which is a monthly measure of the price of a fixed market basket of consumer goods and services. CPI is related to inflation and serves as an economic indicator for government policy.

REFERENCES

Box, G. E. P., and Jenkins, Gwilym M. (1976), *Time Series Analysis Forecasting and Control* (Oakland, CA: Holden Day).

Tukey, John W. (1977), *Exploratory Data Analysis* (Reading, MA: Addison-Wesley).

EXTENDED WRITING ASSIGNMENT

Refer to Appendix 1, "Technical Report Writing," and Appendix 2, "Technical Report Writing Checklist," for guidance on format and style for your report.

Write a report summarizing your findings regarding traffic flow for your chosen location. Keep in mind that you are reporting to a group of nonstatisticians who are not familiar with time series. Your report should include:

1. A description of your chosen location and the data collection method.

2. A graph of the raw data, along with basic summary statistics.

3. A brief description of the smoothing technique and graph(s) of the smoothed data.

4. Any conclusions and recommendations you might have concerning the traffic flow and the timing of the stoplight. (Should the light be changed? Is there heavy, moderate, or very little traffic?)

Name _____ Section _____ Session 6

SHORT ANSWER WRITING ASSIGNMENT

All answers should be complete sentences. Include the raw data plot and a smoothed plot with this assignment.

1. Briefly describe your chosen location. (A sketch indicating where your group observed the car counts may be useful.)

2. Based only on the graph of the original data, what are your initial findings regarding traffic?

3. The mean, median, and standard deviation (respectively) for the vehicle counts are

    $\bar{X} = $ _____ , $M = $ _____ , and $s = $ _____ .

4. Are $\bar{X}$ and the median close to each other? What might this indicate about the shape of the distribution of car counts?

5. Why is it a good idea to smooth some time series plots?

6. Look at your favorite graph of the smoothed car counts. What additional information does it indicate about the car counts for your particular location?

7. In a few sentences, summarize your overall conclusions and any recommendations you have about traffic patterns.

SESSION SEVEN

A Question of Taste

118 SESSION SEVEN

INTRODUCTION

A common experiment in investigating consumer preferences is to give a sample of potential consumers two competing products and ask them which they prefer. The statistical inference involves the proportion of the population of potential consumers who prefer a particular product.

STATISTICAL CONCEPTS

Binomial distribution, proportion, single-blind experiment, randomization.

MATERIALS NEEDED

For each team, two cans of a highly advertised name-brand soft drink, two cans of a nonhighly advertised bargain drink of the same flavor as the name-brand drink, two cups per subject, a cooler and ice or a refrigerator, a calculator, and a coin.

THE SETTING

You are a member of a research team for Check It Out, a market research company. Cheaper Cola wishes to know the proportion of consumers who prefer the taste of their product (the bargain drink) to the competing product (the highly advertised drink) made by King of the Mountain soft drink company. Cheaper Cola has hired Check It Out to perform a consumer taste test and write a report giving the results of the test.

BACKGROUND

Previous research has shown that several factors, in addition to taste, can affect a consumer's reaction to soft drinks. These can include the temperatures of soft drinks, the order in which the drinks are presented, differences in color, the amount and nature of advertising that the soft drinks have received, and the influence of friends and family. Our preference in soft drinks may have as much to do with advertisements we have seen or with what friends like as with the actual taste. It may not be "cool" to drink the bargain drink even if it is very similar in taste to the highly advertised drink.

This particular study deals only with taste preference. We will attempt to control for all other possible effects except order of presentation. We will use random order of presentation so neither soft drink has an advantage and use a **single-blind experiment** in which the subjects are not told the order of presentation of the two drinks.

THE EXPERIMENT

STEP 1: PREPARATION

Make sure the drinks are cooled in an identical fashion for the same length of time so one brand does not get an unfair advantage.

We will eliminate the effects of advertising and peer pressure by not calling the products by their brand names. The products will be presented to the subjects as Brand J and Brand K, rather than by their actual brand names. We will flip a coin to decide which soft drink is called Brand J and which is called Brand K. If the coin is heads, the bargain drink will be called Brand J and the highly advertised drink will be called Brand K. If the coin is tails, the bargain drink will be called Brand K and the highly advertised drink will be Brand J. Write the assignment of brand names in Table 7.1, but do not let the subjects see what you have written.

To avoid giving one brand an undue advantage or disadvantage of always being presented first or always being presented last, we will randomize the order of presentation. For each subject in the experiment, we will flip a coin. If the coin is heads, the subject will be given Brand J first and then Brand K. If the coin is tails, the subject will be given Brand K first and then Brand J. To save time during the performance of the experiment, we will make these assignments in advance. Determine the number of subjects who will be in your experiment. Flip a coin once for each subject. Write the result of the coin flip and the order of brand presentation for each subject in Table 7.2.

For each subject, take two paper cups and lightly write J on the bottom of one cup and K on the bottom of the other cup.

Table 7.1 Assignment of Brand Labels

Brand	Product Name
J	
K	

Table 7.2 Order of Brand Presentation

If coin is heads, give J and then K. If coin is tails, give K and then J.

Subject	Coin (H or T)	1st Brand (J or K)	2nd Brand (J or K)	Subject	Coin (H or T)	1st Brand (J or K)	2nd Brand (J or K)
1				16			
2				17			
3				18			
4				19			
5				20			
6				21			
7				22			
8				23			
9				24			
10				25			
11				26			
12				27			
13				28			
14				29			
15				30			

STEP 2: COLLECTION OF DATA

Warning: Some individuals have serious medical reactions to some ingredients, such as sugar or caffeine, contained in certain soft drinks. Inform all subjects (your classmates) as to the real brand names of the drinks to be used in the study before they are asked to drink the products. If a subject suspects that the product could cause a serious medical reaction, he or she should not participate in the study.

It is important that the subjects do not know the order in which they receive the two drinks. Otherwise, factors other than taste can affect their selection of the preferred taste.

For each subject, the experiment consists of the following:

1. Ask subjects to close their eyes until their part of the experiment is completed.

2. After the subject has his or her eyes closed, pour a small amount (1 ounce) of Brand J and Brand K into cups marked J and K.

3. Using the row of Table 7.2 corresponding to the subject, determine the order in which the subject is to drink the drinks.

4. Instruct the subject to drink a small amount of the first drink and then drink a small amount of the second drink.

5. Have the subject select the drink with the better taste and hand you that cup. The other cup should be discarded. It is permissible for subjects to take more than one drink from each cup to make a decision. It is important for each subject to make a selection.

6. Record the letter on the bottom of the cup next to the subject's number in Table 7.3.

7. Instruct the subject to open his or her eyes.

STEP 3: DATA ANALYSIS

The data analysis for this experiment is based on the numbers of subjects who prefer each brand. Count the number of subjects who prefer Brand J and the number who prefer Brand K. The sum of these two numbers should equal the number of subjects. Write these numbers and the number of subjects in Table 7.4.

Table 7.3 Subject Preference

Subject	Preference (J or K)	Subject	Preference (J or K)	Subject	Preference (J or K)
1		11		21	
2		12		22	
3		13		23	
4		14		24	
5		15		25	
6		16		26	
7		17		27	
8		18		28	
9		19		29	
10		20		30	

Table 7.4 Summary Statistics

Number of subjects preferring Brand J	
Number of subjects preferring Brand K	
Total number of subjects, n	

Number of subjects preferring bargain drink, X	
Sample proportion preferring bargain drink, X/n	
Pr[binomial $(n, 1/2) \leq$ observed value of X]	

Let X denote the number of subjects preferring the taste of the bargain drink (see Table 7.1 to recall whether this is J or K) and n denote the total number of subjects. If we assume that the subjects are a random sample from a large population of potential consumers, then the random variable X has a **binomial distribution with parameters n and p**, where p is the proportion of the population who would prefer the taste of the bargain drink if they were tested in this way. The sample proportion X/n is a point estimate of p. Write X and X/n in Table 7.4.

What does the sample proportion tell you about the taste of the bargain drink relative to the taste of the highly advertised brand? Are you surprised by the results?

If the two drinks have tastes that are equally preferable, then $p = 1/2$—that is, half of the population of potential subjects would prefer the taste of the bargain brand—and X will tend to be close to $n/2$. If a majority of the population favors the highly advertised brand, then $p < 1/2$ and X will tend to be smaller than $n/2$. If a majority of the population favors the bargain brand, then $p > 1/2$ and X will tend to be larger than $n/2$.

How strong is the evidence in support of the notion that the taste of the bargain drink is inferior? To investigate this we can compute the probability that X, the number in a sample of size n who prefer the bargain drink, would be less than or equal to the value we observed if in fact the two drinks are equally preferable. For example, if the drinks are equally preferable and we had $n = 25$ subjects, then the probability that 5 or fewer subjects would prefer the bargain brand is only 0.002039. This very small probability would suggest that if we observed only 5 out of 25 preferring the bargain brand, either the sample was very unusual or the taste of the bargain drink is inferior.

To calculate this probability for our observed value of X, launch Minitab and then:

1. Enter the number of subjects selecting the bargain brand in the first row of column **C1**.

2. Under the **Calc** menu, click and hold on **Probability Distributions** and then select **Binomial** from the submenu. A Binomial Distribution dialog box similar to Figure 7.1 will appear.

3. Click the button to the left of **Cumulative probability**.

4. Type the number of subjects (trials). For example, if $n = 25$, type **25**.

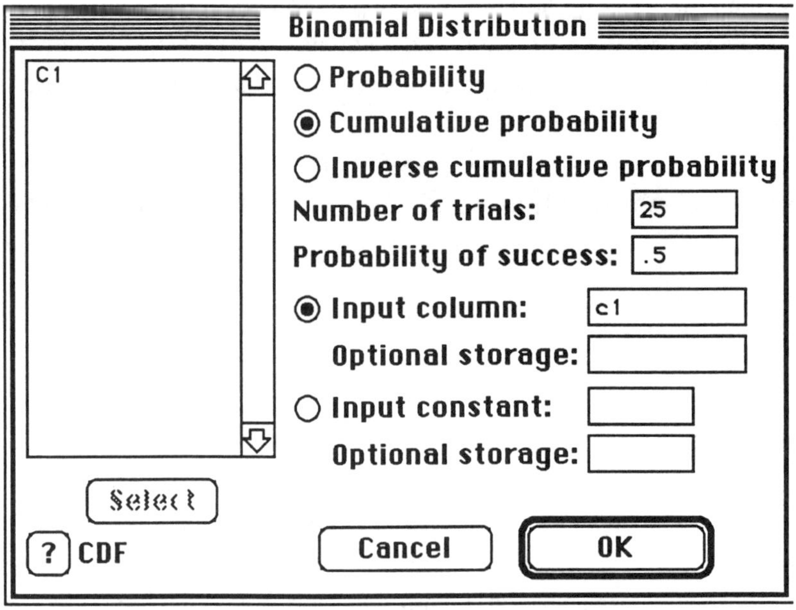

Figure 7.1 Binomial Distribution Dialog Box

5. Click in the box to the right of **Probability of success** and type **0.5**.

6. Click in the box to the right of **Input column** and type **C1**.

7. Click **OK**.

   Minitab now lists the probability that a binomial random variable with parameters n = the number of subjects and p = 1/2 is less than or equal to your observed value of X, the number choosing the bargain drink, in the Session window. Write this probability in Table 7.4 and then quit Minitab by selecting **Quit** under the **File** menu.

   If this probability is very small, there is good reason to doubt the assumption that the drinks are equally preferable (p = 1/2). What does your conclusion tell you about taste preferences among consumers?

If you were to repeat the experiment, what, if anything, would you do differently?

PARTING GLANCES

We have used a single-blind experiment to investigate taste preferences for two soft drinks. We have taken great care to control factors other than taste that might affect consumer preference. Our inference has dealt with p, the proportion of consumers who prefer the taste of the bargain drink. The proportion of our subjects selecting the bargain drink is a point estimate of p.

This was a single-blind experiment because the subject did not know the order in which the drinks were presented but the server knew the order of presentation. In a **double-blind experiment** neither the subject nor the server would know the order of presentation. A double-blind experiment would be more complicated in that a third person would have to prepare the drinks and give them to the server without indicating the brand in each cup. The double-blind experiment has the advantage that the server cannot consciously or subconsciously influence the subject.

A different type of taste-test experiment is known as a triangular taste test. In this experiment, subjects are given three product samples, two of which are identical and a third that is different. Subjects are asked to select the product that is different. If one cannot distinguish the taste differences, then the probability of correctly selecting the different product is 1/3. If one has some ability to distinguish the taste differences then this probability is greater than 1/3.

Many important experiments use the design of exposing a subject to two treatments and observing which treatment is preferred. A medical example is the comparison of two ointments for the treatment of athlete's foot. In such an experiment, a group of subjects having athlete's foot on both feet would be assembled. For each subject, one ointment would be selected at random for use on the subject's left foot and another for use on the right. After a treatment period, a trained observer would judge which foot has the more favorable response. The observer would not be told which ointment was assigned to each foot. The use of both ointments on the same subject controls for subject-to-subject differences that may be a major source of variation.

EXTENDED WRITING ASSIGNMENT

Refer to Appendix 1, "Technical Report Writing," and Appendix 2, "Technical Report Writing Checklist," for guidance on format and style for your report.

Ms. Lisa Marx, marketing manager of Cheaper Cola, wants a report summarizing the experiment. The report should include:

1. A summary of the experiment that you performed.

2. A point estimate for the proportion of consumers who prefer the taste of Cheaper Cola brand to King of the Mountain brand. You should also explain the implications of this estimate to the Cheaper Cola company.

3. A statement of the probability that as few or fewer subjects would prefer the taste of Cheaper Cola brand to King of the Mountain brand if indeed the tastes were equally attractive and its implications to the Cheaper Cola company.

4. A recommendation to Cheaper Cola regarding the need to improve their product or to more heavily advertise the product that they have.

5. A recommendation to Cheaper Cola regarding any changes you would make in the experiment if you could do it again.

Include graphics if you think they would be helpful.

Name _____ Section _____ Session 7

SHORT ANSWER WRITING ASSIGNMENT

All answers should be complete sentences.

1. What feature of this experiment made it a single-blind experiment?

2. Why did we not tell the subjects which drink they were getting first and which drink they were getting second?

3. Complete Table 7.5.

   Table 7.5 Summary Statistics

Number of subjects preferring bargain drink	
Number of subjects preferring high-priced drink	
Total number of subjects	

4. What proportion of your sample preferred the bargain drink?

5. If the two drinks were equally preferable in terms of taste, what is the probability that as few or fewer of the subjects would have selected the bargain drink?

6. Based on your answer to question 5, do you think it is unlikely the drinks are equally preferable in terms of taste? Explain.

7. Explain how a similar experiment could be done to compare the effectiveness of two suntanning lotions.

SESSION EIGHT

Sampling and Variation in Manufactured Products

INTRODUCTION

You might think that all nominal "12 fluid ounce" cans of soft drink contain exactly 12 ounces of liquid or that all nominal "1 inch" screws are exactly 1 inch long. However, this is not true. There is variability among items coming off production lines. There is a great deal of variability in some products and very little in others. Measurements taken on items coming off a production line at a particular point in time follow some unknown probability distribution. This distribution can and often does change over time. For example, the mean or standard deviation may change from one hour to the next.

Much of U.S. industry is being challenged by foreign and domestic competition to improve the quality of their products. The automotive industry is a prime example. The first step in improving quality is to realize that there is variability in manufactured products. One then tries to understand how variable the items are and to identify factors that are producing the variation. These factors may include variability in raw materials and variability in the manufacturing process. By eliminating major sources of variation, a more uniform and generally higher quality product can be produced.

STATISTICAL CONCEPTS

Variation, sources of variation, outliers, descriptive statistics, sampling distribution.

MATERIALS NEEDED

For each team, approximately 200 1/2-inch carpet tacks, a micrometer capable of measuring to 0.001 inch, 2 objects with a premeasured dimension less than 1 inch, and 3 egg cartons with positions numbered from 1 to 36.

MEASURING WITH A MICROMETER

The parts of a micrometer are identified in Figure 8.1. The distance between the anvil and the spindle is adjusted by rotating the thimble and clutch. To make a measurement, place the object of interest between the anvil and the spindle and rotate the clutch until the thimble no longer rotates. The measurement is the sum of three components. The numbered major divisions on the barrel represent distance units of 0.100 inches. The finer divisions on the barrel represent distance measurements of 0.025 inches. The marks on the thimble represent distance units of 0.001 inches. The measurement depicted in Figure 8.1 represents one major division (0.100 inch), one finer division (0.025 inch), and 7 units on the thimble (7 * 0.001 = 0.007 inch), for a total measurement of 0.100 + 0.025 + 0.007 = 0.132 inches.

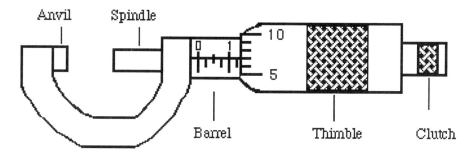

Figure 8.1 Micrometer

Your instructor will provide some objects with known dimensions. Practice measuring these objects before you begin the experiment.

THE SETTING

You are working as a summer intern in your father's company, Yourname Tack Company. Your father has discovered that some of his biggest customers are now buying from a competitor rather than from Yourname. They tell him that Yourname tacks have too much variability. Your tacky father reminds you how much he has spent on your education and tells you it's time for you to produce. He specifically wants you to learn as much as you can about how his tacks vary and to write a report summarizing what you have done and what you have learned.

BACKGROUND

By talking with various plant workers, you discover that a number of measurements can be made on a tack. These include the length, weight, and shaft diameter of the tack, as well as the diameter of the head of the tack, the angle between the shaft and the head, and the sharpness of the point. Variation in these variables can be due to differences in the metal used for the tacks and differences in the production process where the head and point are formed.

Another source of variation in your data has nothing to do with differences among the tacks. This type of variation arises in the process of making the measurements. You may get a slightly different result if you measure the same object twice with the same micrometer. One explanation of this variation is that the micrometer is grasping the object in slightly different places and with slightly different amounts of force in making the two measurements. This type of measurement variation may be larger if the two measurements are made by different people or with different micrometers.

After an interim meeting with your father, you decide to initially study the length of the tack and the diameter of the head. These can all be measured to the nearest 0.001 inch with a micrometer. You have available a sample of tacks that have just come off the production line.

THE EXPERIMENT

STEP 1: OPERATIONAL DEFINITIONS

Before your research team starts measuring tacks, you must precisely define what you are going to measure. By the length of the tack, we mean the distance from the center of the head to the point. The definition of head diameter is trickier. By the diameter of the head, we mean the distance between two points on the outer edge of the head that fall on an imaginary line segment passing through the center of the surface of the head. If the heads of tacks were perfectly round, then the distance would be the same for all such line segments. However, you will notice that the heads are not perfectly round. Your research team may decide to put the head of the tack between the anvil and spindle without paying any attention to the shape of the head. This method approximates a random selection from the set of possible measurements, but there is a real question whether it is really random. Alternatively, your research team may decide to take the micrometer reading with the tack head aligned so as to give the maximum possible diameter measurement. Figure 8.2 illustrates the tack dimensions.

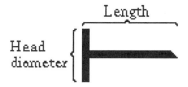

Figure 8.2 Tack Dimensions

In making your decision you should consider the length of time required to make the measurement and the repeatability of the measurement procedure. A repeatable measurement procedure is one that tends to give the same reading on repeated measurements of the same object. The random method will be faster but less repeatable than the maximum method. List your method of measuring head diameter here.

STEP 2: SAMPLE SELECTION

It will not be practical to measure the lengths and head diameters of all tacks coming off the production line. Your research team has been given a large number of tacks. You are to randomly select 36 of these for measurement with the micrometer. You should do this in a manner that approximates a simple random sample. This can be done by spreading the tacks out on a surface and randomly selecting tacks from a wide variety of locations with your eyes closed. *Be careful to pick up any tacks that fall off the table.* After you have selected your sample of size 36, put the remaining tacks back in the container.

STEP 3: COLLECTION OF THE DATA

Measure and record in Table 8.1 the length and head diameter of each tack and place each tack in the egg carton location corresponding to its number. The measurements can be made quicker by first measuring all of the lengths and then measuring all of the head diameters.

Table 8.1 Length and Head Diameter of Tacks

Tack	Length	Diameter	Tack	Length	Diameter
1			19		
2			20		
3			21		
4			22		
5			23		
6			24		
7			25		
8			26		
9			27		
10			28		
11			29		
12			30		
13			31		
14			32		
15			33		
16			34		
17			35		
18			36		

STEP 4: DATA ANALYSIS

Choose a Macintosh and launch Minitab. Before entering your tack data, name the first two columns **Length** and **Diameter**. The Untitled worksheet should now look like Figure 8.3. Carefully enter the data from Table 8.1 in the worksheet, putting the measured lengths of your tacks in column C1 and the diameters in column C2.

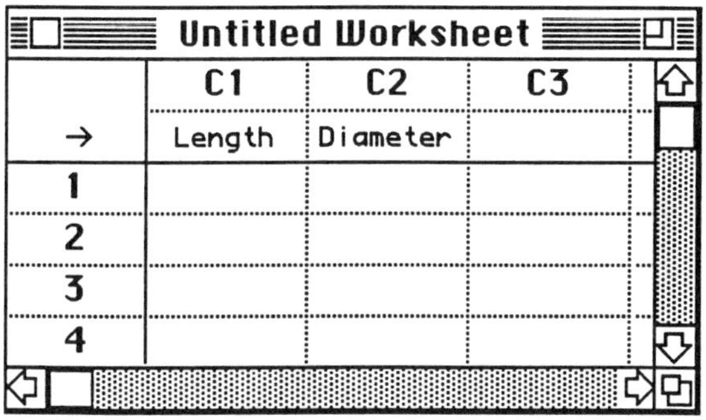

SAMPLING AND VARIATION IN MANUFACTURED PRODUCTS   135

Figure 8.3 Untitled Worksheet

After entering the data for each tack, go through the worksheet and make sure that all of the data has been entered correctly. The computer cannot give a correct analysis if you entered incorrect data. When you are sure the data is correct, save it on your diskette as **Tackdata** by selecting **Save Worksheet As** from the **File** menu.

Let's begin our data analysis by looking at some descriptive statistics for both variables. To do this:

1. Under the **Stat** menu, click and hold on **Basic Statistics** and then select **Descriptive Statistics** from the submenu. A Descriptive Statistics dialog box similar to Figure 8.4 will appear.

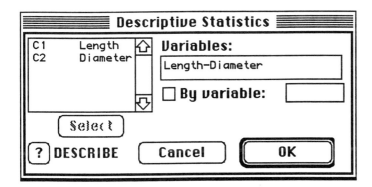

Figure 8.4 Descriptive Statistics Dialog Box

2. Double-click on the variable names **Length** and **Diameter**.

3. Click **OK**.

136   SESSION EIGHT

Table 8.2 Descriptive Statistics

Variable	Sample Mean	Sample Standard Deviation
Length		
Diameter		

Separate sets of descriptive statistics for the variables Length and Diameter will now appear in the Session window. It is good practice to check the reported number of observations and the minimum and maximum values for each variable. Are they what you expected them to be? If not, you may have entered some of the data incorrectly. For example, the incorrect placement of a decimal point can often be detected by looking at minimums and maximums. If you want to check your data again, close the Session window and your worksheet will appear. If you have errors in your data set, correct them, save the changes by selecting **Save Worksheet** from the **File** menu, and produce a new set of descriptive statistics for these variables.

After you are sure that your data has been entered correctly, record the sample mean and sample standard deviation for each variable in Table 8.2. Do the tacks exhibit variability in their lengths and diameters?

Let's now learn about the distribution of each variable by producing stem-and-leaf diagrams. To do this:

1. Under the **Graph** menu, select **Stem-and-Leaf**. A Stem-and-Leaf dialog box similar to Figure 8.5 will appear.

2. Double-click on the variable names **Length** and **Diameter**.

3. Click **OK**.

Separate stem-and-leaf diagrams for the variables Length and Diameter will now appear in the Session window and as output on the printer. Note that you also could have obtained histograms and dotplots for the variables through the Graph menu.

Look at your stem-and-leaf diagrams. Do you see any observations that are very different from the rest? Statisticians call such observations **outliers**. There are two

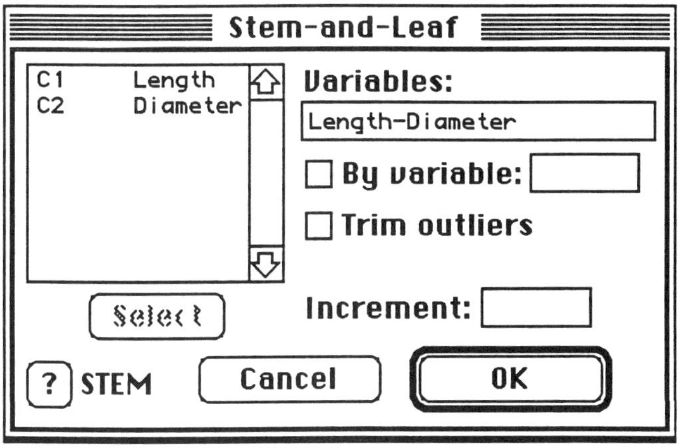

Figure 8.5 Stem-and-Leaf Dialog Box

reasons for outliers. Either you made a big mistake in making the measurement or the tack is very different from the rest of the tacks.

If you find outliers, use Table 8.1 to identify the tack number and redo the measurement. If you made a big mistake in your previous measurement, correct the error in the data worksheet, save the changes, and redo the descriptive statistics and stem-and-leaf diagrams. If you did not make a big mistake, then the tack is just different from the rest.

Based on your stem-and-leaf diagrams, how would you describe the shape of the distributions for the two variables? Are they approximately normal? Are they skewed in one direction or the other? Is there more than one mode? What feature(s) of production might cause more than one mode in a distribution?

138   SESSION EIGHT

Let's now look at the relationship, if any, between the variables Length and Diameter. We will produce a plot with Length on the vertical axis and Diameter on the horizontal axis. To do this:

1. Under the **Graph** menu, select **Scatter Plot**. A Scatter Plot dialog box similar to Figure 8.6 will appear.

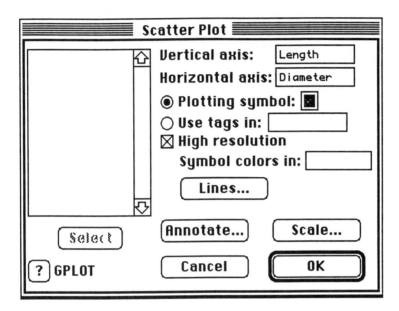

Figure 8.6 Scatter Plot Dialog Box

2. Click in the box next to **Vertical axis** and type **Length**.

3. Click in the box next to **Horizontal axis** and type **Diameter**.

4. Click **OK**.

After a short delay, a scatter plot will appear in the Gplot window. Print the plot now by selecting **Print Window** under the **File** menu. What, if any, patterns do you see in the scatter plot?

Return to the Data window by selecting **Data** under the **Window** menu.

Statistical theory suggests that the sample means of groups of observations tend to be less variable than the individual observations. To investigate this idea, let's treat our set of 36 tack measurements in Table 8.1 as 12 samples of size three. Scroll to the top of the Data window. Enter the number 1 in column C3 for the first three tacks, enter the number 2 for the next three tacks, and so on. After you have entered 12 for the last three tacks, you are ready to compute descriptive statistics for the samples of size three. To compute separate descriptive statistics for each of the 12 samples of size three, we will use Minitab's **by-variable** feature.

1. Under the **Stat** menu, click and hold on **Basic Statistics** and then select **Descriptive Statistics** in the submenu. A Descriptive Statistics dialog box similar to Figure 8.7 will appear.

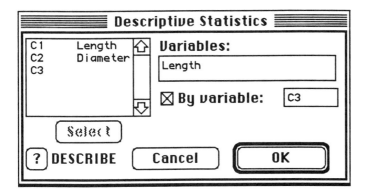

Figure 8.7 Descriptive Statistics Dialog Box

2. Double-click on the variable name **Length**.
3. Click in the box to the left of **By variable**.
4. Click in the box to the right of **By variable** and type **C3**.
5. Click **OK**.

For each sample of size three, separate descriptive statistics for the variable Length will appear in the Session window. Write the sample mean for each sample in Table 8.3. Is there variability among the sample means?

140   SESSION EIGHT

Table 8.3 Sample Means for Samples of Size 3

Sample	1	2	3	4	5	6
Mean						

Sample	7	8	9	10	11	12
Mean						

Return to the worksheet and enter the 12 sample means in the first 12 rows of column C4. Let's now learn about the distribution of the sample means.

1. Under the **Graph** menu, select **Stem-and-Leaf**. A Stem-and-Leaf dialog box similar to Figure 8.8 will appear.

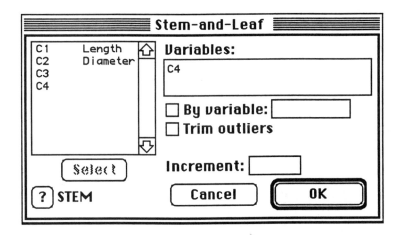

Figure 8.8 Stem-and Leaf Dialog Box

2. Double-click on **C4**.

3. Click **OK**.

Does the stem-and-leaf diagram for these sample means look more or less bell-shaped than the original stem-and-leaf diagram for the variable Length?

To find the amount of variability among the sample means:

1. Under the **Stat** menu, click and hold on **Basic Statistics** and then select **Descriptive Statistics** in the submenu. A Descriptive Statistics dialog box similar to Figure 8.9 will appear.

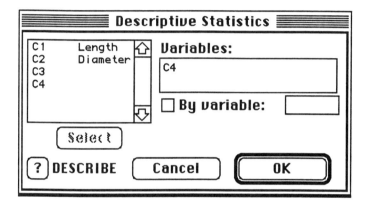

Figure 8.9 Descriptive Statistics Dialog Box

2. Double-click on the variable **C4**.
3. Click **OK**.

The descriptive statistics that appear in the Session window are for the data set of 12 sample means. What is the value of the sample standard deviation of this data set and how does it compare with the standard deviation for the variable Length, which is given in Table 8.2? Does this agree with statistical theory?

This concludes our calculations on the data. Print your Session window by selecting **Print Window** under the **File** menu. Make sure you have all the output you need, and then quit Minitab by selecting **Quit** from the **File** menu.

PARTING GLANCES

We have studied the variability in measurements on a selection of tacks. By looking at the sample standard deviations and at the stem-and-leaf diagrams, we have an idea of the variation of the measurements. The stem-and-leaf diagrams also give us a picture of the distribution of the measurements. The scatter plot allowed us to explore for relationships between the length and head diameter of the tacks.

The variation in our measurements is composed of variation due to true differences among the tacks and variation due to measurement error. Our experiment did not attempt to separate these two types of variation. A different type of experiment where each tack was measured several times would allow one to separate these sources of variation. This type of experiment was not done here because it is more time-consuming and the analysis is more complicated. By using precision measuring equipment and well-trained workers, the variation due to measurement error can be greatly reduced.

Elimination of variation leads to a higher quality product. Consider two parts that must fit together, such as a piston and a cylinder. If there is great variation in the part diameters, the two do not work well together in their common job of transferring energy. If the variation is too great, then the parts may not fit together at all and must be scrapped. The result is a lower quality, higher cost engine.

EXTENDED WRITING ASSIGNMENT

Refer to Appendix 1, "Technical Report Writing," and Appendix 2, "Technical Report Writing Checklist," for guidance on format and style for your report.

This is your chance to show your father that you deserve a promotion to vice-president, a fancy office, and an expense account. Write a report that summarizes what you have done. The report should include:

1. A description of the experiment you performed including operational definitions for the variables you measured.

2. A description of the distribution of the two variables you measured. This description should refer to, at least, the shape of the distribution, the center of the distribution, and the amount of variability. Your description should include both text and graphics.

3. A description of the relationship, if any, between the two variables. Again, include both text and graphics.

4. A comparison of the distribution of the variable Length and the sampling distribution of the sample mean of Length with $n = 3$.

5. Any recommendations for action within the factory based on your results.

6. Any recommendations for further experimentation based on your results.

Name   Section   Session 8

SHORT ANSWER WRITING ASSIGNMENT

All answers should be complete sentences. Include a copy of your stem-and-leaf diagrams for length, head diameter, and C4 and your scatter plot for length versus head diameter with this assignment.

1. What was your operational definition for the length of a tack? Be specific enough so everyone will know how to make the measurement without asking any questions.

2. What was your operational definition for the head diameter of a tack? Why did your group decide on this definition?

3. Did you observe any outliers in your stem-and-leaf diagrams? If so, were these outliers the result of measurement errors or true differences among the tacks?

4. How would you describe the distribution of the length of the tacks (e.g., bell shaped, symmetric, skewed, bimodal)? Also give the mean and the standard deviation.

5. How would you describe the distribution of the head diameter of the tacks (e.g., bell shaped, symmetric, skewed, bimodal)? Also give the mean and the standard deviation.

6. What relationship, if any, did you discover between the length and the head diameter variables of the tacks based on the scatter plot? Are there any patterns?

7. Are the sample means given in Table 8.3 more variable or less variable than the individual lengths given in Table 8.1? What is expected from statistical theory about the sampling distribution of the sample mean?

SESSION NINE

Exploring Statistical Theory Through Computer Simulation

INTRODUCTION

In a typical experiment, we select a sample, make some measurements, and compute a statistic. We might select a sample of students, measure their heights, and compute the sample mean. This sample mean serves as an estimate of the population mean.

In doing such an experiment you might ask yourself some questions, such as, "If I repeated the experiment several times with different samples, would I get different values for my sample mean?" To answer this question, consider computing the average height for all possible samples of three students selected from all students at your school. Unless all of the students in your school are exactly the same height, the sample means will not be the same for all samples. The distribution of the values of a statistic across all possible samples is known as the statistic's **sampling distribution**. The sampling distribution of a statistic generally depends on the sample size and on the distribution of the original measurements for the members of the population. Thus, the sampling distribution of the sample mean depends on the number of students included in your sample and on the distribution of heights for students in your school.

Additional questions might be, "Why did I compute the sample mean? Would it have been better to have computed the sample median or some other statistic?" To answer this question we need to investigate and compare the sampling distributions of the proposed statistics.

Statisticians investigate sampling distributions of statistics in two ways. In the first approach, mathematical arguments are used to determine the sampling distribution. These arguments can be quite complicated and are beyond the scope of this laboratory. You may be surprised to learn that for some statistics, no statistician has yet been able to mathematically derive the sampling distribution!

The second approach for investigating sampling distributions involves using a computer to simulate the selection of a large number of samples from a population. In this laboratory, we will demonstrate simulation using Minitab on the Macintosh. In actual simulation studies, statisticians would use a more flexible programming language such as FORTRAN, C, or Pascal.

STATISTICAL CONCEPTS

Sampling distributions, comparison of estimators, sample mean, sample median, central limit theorem, simulation.

MATERIALS NEEDED

None.

THE SETTING

You are a statistics professor at a major university. In your current research project, you are trying to learn more about the sampling distributions of estimators of the center of a population. This knowledge will allow users of statistics to gain more information from their data.

BACKGROUND

There are several ways to measure the center of a sample or population. Two popular measures are the **mean** and the **median**. The sample mean is the sum of the responses in the sample divided by the sample size. Hence, it is the average response in the sample. The population mean is the average response in the population.

The sample median is a number such that at least half of the responses in the sample are greater than or equal to the number and at least half of the responses in the sample are less than or equal to the number. Hence it is a midpoint of the ordered sample. To compute the median by hand, we order the observations from smallest to largest. If our sample size is odd, then the sample median is the middle observation in the ordered sample. If the sample size is even, the sample median is the average of the two middle-most scores in the ordered sample. The population median is the midpoint of the population.

If we sample from a **symmetric** population, then the population mean and the population median are equal. In this case, the sample mean and the sample median are estimating the same population parameter. We could use either the sample mean or the sample median to estimate this unknown parameter. It is not clear which would be the better estimator. It may be that one is better for all populations. On the other hand, perhaps the sample mean is better for some types of populations and the sample median is better for other types. Our laboratory session will give us some information for comparing the sampling distributions of these two statistics.

If we sample from a **nonsymmetric** population, then the population mean and the population median are generally unequal. The sample mean and the sample median are estimating different population parameters. Hence the sample mean and the sample median are not directly comparable for nonsymmetric populations. Figure 9.1 illustrates three symmetric populations and a nonsymmetric population.

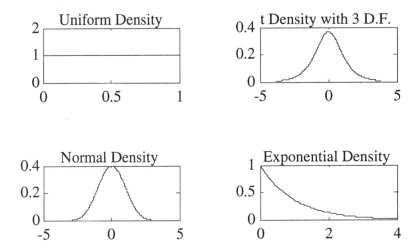

Figure 9.1 Uniform, t, Normal, and Exponential Density Functions

THE EXPERIMENT

STEP 1: SIMULATION OF SAMPLING FROM A UNIFORM DISTRIBUTION

We begin this step by using Minitab to generate 600 samples of size 20 from the uniform distribution on the (0, 1) interval of the real line. The density curve for this distribution is illustrated in Figure 9.1. This uniform distribution is symmetric about the point 0.5, its mean and median. These computer-generated observations will appear in the worksheet as 600 rows and 20 columns. Thus, each row will represent a different sample of size 20. The first column will represent the first observation in all 600 samples.

To generate the observations from the uniform distribution, choose a computer and launch Minitab.

1. Under the **Calc** menu, click and hold on **Random Data** and then select **Uniform** in the submenu. This causes a Uniform Distribution dialog box similar to Figure 9.2 to appear.

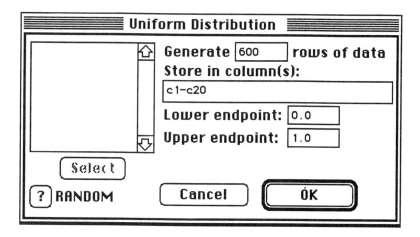

Figure 9.2 Uniform Distribution Dialog Box

2. Click in the box to the right of **Generate** and type **600**.

3. Click in the box under **Store in column(s)** and type **C1-C20.**

4. Click **OK**.

After a considerable period of time, 600 rows of numbers between 0 and 1 will appear in columns C1, C2, ..., C20 of the worksheet. These are the 600 simulated samples of size 20 from the uniform distribution. We will use these 600 samples to learn about the sampling distribution of the sample mean and the sample median.

We will now compute the sample mean for each row and place the result in C21.

1. Under the **Calc** menu, click and hold on **Functions and Statistics** and then select **Row Statistics** in the submenu. This causes a Row Statistics dialog box similar to Figure 9.3 to appear.

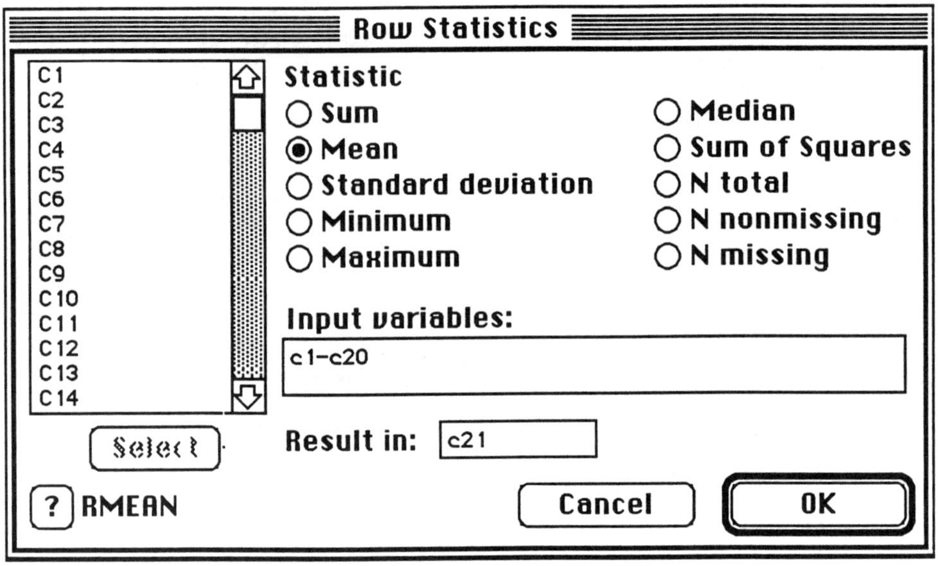

Figure 9.3 Row Statistics Dialog Box

2. Click in the circle to the left of **Mean** in the Statistic list.

3. Click in the box under **Input variables** and type **C1-C20**.

4. Click in the box to the right of **Results in** and type **C21**.

5. Click **OK**.

After a short delay, Minitab adds column C21 to the worksheet. The column contains the sample mean of the items in columns 1 through 20 for each row. Thus, C21 contains the sample means for our 600 computer-generated samples of size 20 from the uniform distribution. If we had clicked on Median rather than Mean in the list of statistics in the Row Statistics dialog box, we would have computed the sample median for each row. Modify the above steps and have Minitab compute the sample median for each computer-generated sample and place the results in column C22.

After Minitab completes the calculations for C22, scroll to the top of the worksheet and name C21 **Umean** and C22 **Umedian** to indicate that the columns represent means and medians from a uniform distribution. After naming the variables you must use the mouse or the arrows on the keyboard to move the pointer into the data.

We are now ready to investigate the sampling distribution of the sample mean and the sample median. We will do this by producing histograms and computing descriptive statistics for the sets of 600 means and 600 medians.

To produce histograms:

1. Under the **Graph** menu, select **Histogram**. This causes a Histogram dialog box similar to Figure 9.4 to appear.

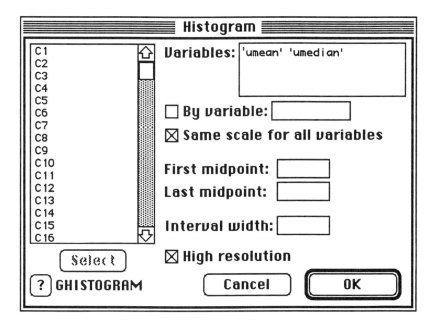

Figure 9.4 Histogram Dialog Box

2. Click in the box to the right of **Variables** and type **'Umean' 'Umedian'**.

3. Click in the box to the left of **Same scale for all variables**.

4. Click **OK**.

A window containing the histogram for Umean will appear, and then a window containing the histogram for Umedian will appear on top of the first window. You will need a printed copy of these histograms for your report. We will print the histogram for Umedian and then the histogram for Umean. To print the histogram for Umedian, select **Print Window** under the **File** menu. *Note:* The draft mode on ImageWriter II printers will not print the histogram.

After a short delay, the histogram for Umedian will be printed. We are now ready to print the histogram for Umean. Close the window containing the histogram for Umedian and repeat the steps that you used to print the preceding histogram. These histograms approximate the sampling distributions of the sample mean and the sample median.

To produce descriptive statistics for these sets of sample means and sample medians:

1. Under the **Stat** menu, click and hold on **Basic Statistics** and then select **Descriptive Statistics** in the submenu. This causes a Descriptive Statistics dialog box similar to Figure 9.5 to appear.

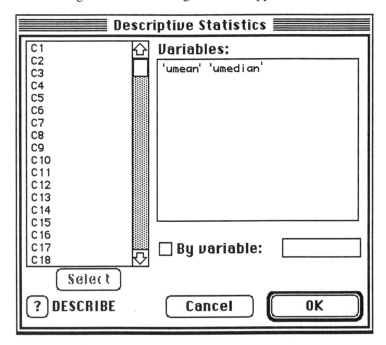

Figure 9.5 Descriptive Statistics Dialog Box

2. Click in the box under **Variables** and type **'Umean' 'Umedian'**.

3. Click **OK**.

The descriptive statistics that appear in the Session window summarize the data sets of 600 sample means and 600 sample medians. For example, the entry in the column Mean for the variable Umean is the sample mean for the set of 600 sample

Table 9.1 Descriptive Statistics Summarizing 600 Values of Sample Mean and Sample Median Computed on Samples of Size 20

Population	Sample Mean		Sample Median	
	Mean	Standard Deviation	Mean	Standard Deviation
Uniform (Step 1) Pop. mean = 0.5 Pop. median = 0.5				
t 3 d.f. (Step 2) Pop. mean = 0.0 Pop. median = 0.0				
Normal (Step 3) Pop. mean = 0.0 Pop. median = 0.0				
Exponential (Step 4) Pop. mean = 1.0 Pop. median = 0.693				

means. The entry in the column StDev for the variable Umean is the sample standard deviation for the set of 600 sample means. The corresponding entries for the variable Umedian are the sample mean and sample standard deviation for the set of 600 sample medians. Write the sample mean and sample standard deviation for the variables Umean and Umedian in the row of Table 9.1 corresponding to the uniform population.

Recall that the population mean and the population median of this uniform distribution are both equal to 0.5. Thus, we would like to estimate them with a statistic whose average value is very close to 0.5 and whose standard deviation is as small as possible.

Use the histograms and the descriptive statistics to answer the following questions.

How would you describe the shape of the sampling distribution of the sample mean for samples of size 20 from a uniform distribution? Is it bell shaped?

How would you describe the shape of the sampling distribution of the sample median for samples of size 20 from a uniform distribution? Is it bell shaped?

Does the center of the sampling distribution of the sample mean appear to be close to the desired value of 0.5?

Does the center of the sampling distribution of the sample median appear to be close to the desired value of 0.5?

Does one of the sampling distributions appear to have more variability than the other? Explain.

Before moving to Step 2, save the Minitab worksheet on your diskette by selecting **Save Worksheet As** under the **File** menu. Name the worksheet **Uniform**. Close the Data window and then close the Session window.

STEP 2: SIMULATION OF SAMPLING FROM THE T DISTRIBUTION WITH THREE DEGREES OF FREEDOM

There is a family of probability distributions known as t distributions. Members of the family are distinguished by their number of degrees of freedom (d.f.). We will be investigating the sampling distribution of the sample mean and the sample median when sampling from the t distribution with three degrees of freedom. The density curve for this distribution is illustrated in Figure 9.1. The distribution is symmetric about the point 0, its mean and median. The possible values of the t variable are all points on the real line. The t distributions have less density in the center and more density in the tails than the standard normal distribution.

To save time, 600 samples of size 20 from the t distribution with 3 degrees of freedom have already been generated. To access this data:

1. Under the **File** menu, select **Open Worksheet**.
2. Double-click on the **Simulation** folder.
3. Double-click on the **t 3df simulation** file.

After a short delay a worksheet will appear with 600 rows and 22 columns. The 600 rows represent 600 simulated samples. The 20 observations from the samples appear in C1 to C20. The sample mean for each sample appears in C21. This column is named Tmean. The sample median for each sample appears in C22. This column is named Tmedian.

Use Minitab to construct histograms and to compute descriptive statistics for the variables Tmean and Tmedian as was done in Step 1 for Umean and Umedian. Write the sample mean and the sample standard deviation for the variables Tmean and Tmedian in the row of Table 9.1 corresponding to the t population. Print the histograms. This t distribution has population mean and population median equal to zero. We would like to estimate them with a statistic whose average value is very close to zero and whose standard deviation is as small as possible.

Use the histograms and descriptive statistics for Tmean and Tmedian to answer the following questions.

How would you describe the shape of the sampling distribution of the sample mean for samples of size 20 from a t distribution with three degrees of freedom? Is it bell shaped?

How would you describe the shape of the sampling distribution of the sample median for samples of size 20 from a t distribution with three degrees of freedom? Is it bell shaped?

Does the center of the sampling distribution of the sample mean appear to be close to the desired value of zero?

Does the center of the sampling distribution of the sample median appear to be close to the desired value of zero?

Does one of the sampling distributions appear to have more variability than the other? Explain. Is this pattern similar to or different from what you observed for the uniform distribution?

Before moving to Step 3, close the Data window and close the Session window.

STEP 3: SIMULATION OF SAMPLING FROM THE STANDARD NORMAL DISTRIBUTION

We now investigate the sampling distribution of the sample mean and the sample median when sampling from the standard normal distribution. The density curve for this distribution is illustrated in Figure 9.1. The distribution is symmetric about the point 0, its mean and median. The possible values of standard normal variables are all points on the real line.

To save time, 600 samples of size 20 from the standard normal distribution have already been generated. To access this data:

1. Under the **File** menu, select **Open Worksheet**.

2. Double-click on the **Simulation** folder.

3. Double-click on the **normal simulation** file.

After a short delay a worksheet will appear with 600 rows and 22 columns. The 600 rows represent 600 simulated samples. The 20 observations from the samples appear in C1 to C20. The sample mean for each sample appears in C21. This column is named Nmean. The sample median for each sample appears in C22. This column is named Nmedian.

Use Minitab to construct histograms and to compute descriptive statistics for the variables Nmean and Nmedian. Write the sample mean and the sample standard deviation for the variables Nmean and Nmedian in the row of Table 9.1 corresponding to the normal population. Print the histograms. This normal distribution has population mean and population median equal to zero. We would like to estimate them with a statistic whose average value is very close to zero and whose standard deviation is as small as possible.

Use the histograms and descriptive statistics for Nmean and Nmedian to answer the following questions.

How would you describe the shape of the sampling distribution of the sample mean for samples of size 20 from the standard normal distribution? Is it bell shaped?

How would you describe the shape of the sampling distribution of the sample median for samples of size 20 from the standard normal distribution? Is it bell shaped?

Does the center of the sampling distribution of the sample mean appear to be close to the desired value of zero?

Does the center of the sampling distribution of the sample median appear to be close to the desired value of zero?

Does one of the sampling distributions appear to have more variability than the other? Explain.

Before moving to Step 4, close the Data window and close the Session window.

STEP 4: SIMULATION OF SAMPLING FROM THE EXPONENTIAL DISTRIBUTION

We now investigate the sampling distribution of the sample mean and the sample median when sampling from the exponential distribution. The density curve for this distribution is illustrated in Figure 9.1. Unlike the other distributions we have considered, the exponential distribution is nonsymmetric. The density has a very long right tail. The population mean is 1 and the population median is approximately 0.693. The possible values of exponential variables are all positive numbers.

To save time, 600 samples of size 20 from the exponential distribution with population mean equal to one have already been generated. To access this data:

1. Under the **File** menu, select **Open Worksheet**.
2. Double-click on the **Simulation** folder.
3. Double-click on the **exponential simulation** file.

After a short delay a worksheet will appear with 600 rows and 22 columns. The 600 rows represent 600 simulated samples. The 20 observations from the samples appear in C1 to C20. The sample mean for each sample of size 20 appears in C21. This column is named Emean. The sample median for each sample of size 20 appears in C22. This column is named Emedian.

Use Minitab to construct histograms and to compute descriptive statistics for the variables Emean and Emedian. Write the sample mean and the sample standard deviation for the variables Emean and Emedian in the row of Table 9.1 corresponding to the exponential population. Print the histograms. We would like to estimate the population mean with a statistic whose average value is very close to 1 and whose standard deviation is as small as possible. We would like to estimate the population median with a statistic whose average value is very close to 0.693 and whose standard deviation is as small as possible.

Use the histograms and descriptive statistics for Emean and Emedian to answer the following questions.

How would you describe the shape of the sampling distribution of the sample mean for samples of size 20 from the exponential distribution? Is it bell shaped?

How would you describe the shape of the sampling distribution of the sample median for samples of size 20 from the exponential distribution? Is it bell shaped?

Does the center of the sampling distribution of the sample mean appear to be close to the desired value of one?

Does the center of the sampling distribution of the sample median appear to be close to the desired value of 0.693?

Quit Minitab by selecting **Quit** under the **File** menu.

STEP 5: COMPARISON WITH THEORY

In this step we compare your results for all distributions and compare them with statistical theory. Statistical theory says that the mean of the sampling distribution of the sample mean equals the population mean. Was the center of the sampling distribution of the sample mean close to the population mean for each distribution in your study?

The central limit theorem says that for large samples from a population with mean μ and variance σ^2, the sampling distribution of the sample mean is approximately a normal distribution with mean μ and variance σ^2/n. Moreover, statistical theory says that when sampling from a normal population, the sampling distribution of the sample mean is normal for all sample sizes. In your study, did you find the sampling distributions of the sample mean to be approximately normal for samples of size 20?

The sampling distribution of the sample mean can be far from normal for small sample sizes. Figure 9.6 gives a histogram for a set of 600 sample means computed from samples of size 2 from the exponential distribution considered in Step 4. Compare the histogram in Figure 9.6 with your histogram for Emean with regard to the shape and variability of the sampling distributions. Note that the axis scales are different in the two histograms.

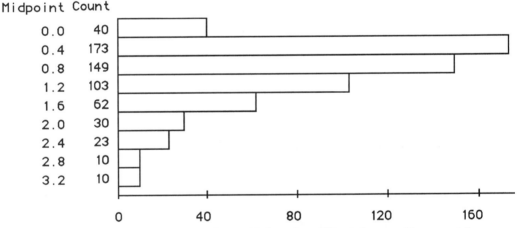

Figure 9.6 Histogram of Sample Means from 600 Samples of Size 2 from an Exponential Distribution

Based on your studies with the uniform, t, and normal distributions, is the sample mean always the best statistic for estimating the common mean and median of a symmetric distribution?

Most elementary statistics books do not discuss the sampling distribution for the sample median. Based on your histograms, do you suspect that the sampling distribution of the sample median is approximately a normal distribution for large sample sizes?

PARTING GLANCES

We have investigated the sampling distributions of the sample mean and the sample median by having the computer simulate random samples from four different distributions. The histograms produced from the 600 sample means and 600 sample medians are approximations of the sampling distributions of the sample mean and sample median. When sampling from symmetric distributions, the sample mean and the sample median are estimating the same parameter. For some symmetric distributions, the sample mean is a better estimator, while for others the sample median is better. When sampling from nonsymmetric distributions, the sample mean and the sample median estimate different parameters and are not directly comparable.

The central limit theorem states that the sampling distribution of the sample mean is approximately normal for sufficiently large sample sizes. By looking at the shape of the histograms for the sample means, we were able to see how good the approximation is for samples of size 20.

We have had the computer generate random numbers to compare two estimators. Scientists also use random numbers to model complicated processes, such as to model traffic flow in a large city.

EXTENDED WRITING ASSIGNMENT

Refer to Appendix 1, "Technical Report Writing," and Appendix 2, "Technical Report Writing Checklist," for guidance on format and style for your report.

Write a report summarizing the experiment. It should include:

1. A description of the simulation experiment that you performed

2. A graphic and numerical description of the results of your experiments for the four distributions

3. A description of how your results compare with the central limit theorem

4. Your recommendations for when to use the sample mean and when to use the sample median

Name _____ Section _____ Session 9

SHORT ANSWER WRITING ASSIGNMENT

All answers should be complete sentences. Include a copy of Table 9.1 with this assignment.

1. List the two approaches statisticians use to investigate the sampling distributions of statistics.

2. For the symmetric distributions that we studied, the expected values of the sample mean and the sample median are identical. Thus, we wish to use the statistic whose sampling distribution has the smaller standard deviation. Indicate the better estimator for the following symmetric distributions:

   Uniform   ___ sample mean ___ sample median ___ both are equally good
   t with 3 d.f. ___ sample mean ___ sample median ___ both are equally good
   Normal   ___ sample mean ___ sample median ___ both are equally good

3. Is either statistic better than the other for *all* symmetric distributions? Explain.

4. Based on Figure 9.6, is the sampling distribution of the sample mean well approximated by the normal distribution for samples of size 2 from an exponential distribution? Explain.

5. Based on your histogram of the 600 sample means for samples of size 20 from the exponential distribution, is the sampling distribution of the sample mean closer to being normal for samples of size 20 than it is for samples of size 2?

6. Based on your histograms for the sample medians, do you think a theorem like the central limit theorem is true for the sample median? Explain.

SESSION TEN

Improving Product Performance with Planned Experiments

INTRODUCTION

Researchers use planned experiments to study the effects of factors on the mean performance of a product or process. By performing planned experiments they can determine which factors affect mean performance. In many cases the results of planned experiments are used to select levels of each factor in an attempt to optimize mean performance.

STATISTICAL CONCEPTS

Planned experiment, two-factor design, factor selection, randomization, interaction.

MATERIALS NEEDED

For each team, two balsa wood or Styrofoam airplanes, a 50-foot tape measure, a straightedge, a paper clip, a calculator, and scissors.

THE SETTING

You are a member of a research team in the Product Design and Development Division of Better Balsa Aerospace Corporation. Your research team is attempting to improve flight performance of the Model XJ2 glider aircraft. Your objective is to design the product to maximize the mean distance of flight. Budget and time constraints will allow for a maximum of 16 test flights in your investigation.

BACKGROUND

Previous experience with similar aircraft suggests that several factors can affect the mean distance of flight. These include uncontrollable factors such as wind and humidity and controllable factors such as the size of the wings, the shape of the wings, the presence or absence of extra weight on the nose, the size of the tail, the shape of the tail, the angle of the wings relative to the ground at launch, the height off the ground at launch, and the amount of force used in the launch.

THE EXPERIMENT

STEP 1: CHOICE OF FACTORS AND LEVELS

Your research team is to choose two controllable factors (e.g., angle of the wings relative to the ground at launch and wing size) for investigation in the study. We denote these factors by factor A and factor B. For each selected factor, choose two settings or levels. We denote the levels by level 1 and level 2. For example, if factor A is the angle of wing, level 1 could represent 0 degrees and level 2 could represent 20 degrees. If factor B is wing size, level 1 could represent the current wing size and level 2 could represent a 20 percent reduction in wing size. Write your factors and their levels in Table 10.1.

In performing the test flights, attempt to hold all potential factors constant except for factors A and B. For example, we could attempt to keep the force of launch constant by having the same person launch all flights and instructing him or her to launch the planes as consistently as possible. By controlling all other potential factors, we get a true measure of the effects of the factors we are varying in the experiment.

For each test flight, you will use a level of factor A and a level of factor B. Statisticians refer to the combination of a level of factor A and a level of factor B as a **treatment combination**. This experiment has four treatment combinations:

(1, 1) Level 1 of factor A and level 1 of factor B

(1, 2) Level 1 of factor A and level 2 of factor B

(2, 1) Level 2 of factor A and level 1 of factor B

(2, 2) Level 2 of factor A and level 2 of factor B

As 16 test flights are possible, we will have 4 test flights for each treatment combination. This is an example of a **two-factor factorial experiment with replication**.

STEP 2: RANDOMIZATION

While we attempt to control all factors in our experiment, there are some factors that are either beyond our control, such as wind currents, or that we are not aware of at this stage of product development. To guard against one treatment combination getting an unfair advantage, we will use **randomization**. That is, we will use random numbers to decide which 4 of the 16 flights will be assigned to each treatment combination. In this way all treatment combinations have an equal chance of being assigned to all sets of 4 flights and no treatment combination has any advantage.

We will use Minitab to generate random numbers for use in randomizing the assignment of treatment combinations to the flights. These random numbers have the property that the numbers 1 and 2 are equally likely to occur at each point in the sequence. To generate these random numbers, launch Minitab and do the following:

1. Under the **Calc** menu, click and hold on **Random Data** and then select **Integer** from the submenu. An Integer Distribution dialog box similar to Figure 10.1 will appear.

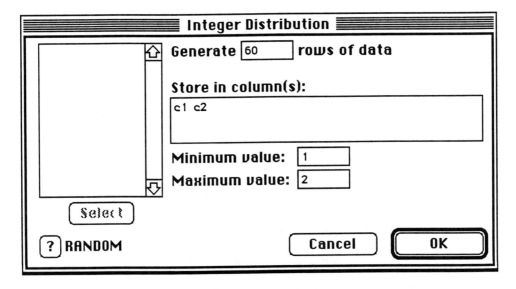

Figure 10.1 Integer Distribution Dialog Box

2. Click in the box next to **Generate** and type the number of random numbers you want. For this experiment 60 random numbers should be sufficient.

3. Click in the box beneath **Store in column(s)** and type **C1 C2**.

4. Click in the box to the right of **Minimum value** and type **1**.

5. Click in the box to the right of **Maximum value** and type **2**.

6. Click **OK**.

Table 10.1 List of Factors and Their Levels

Factor A	
Level 1	
Level 2	

Factor B	
Level 1	
Level 2	

Table 10.2 Flight Schedule by Treatment Combination

Flight Number	Treatment Combination	Flight Length
	(1, 1)	
	(1, 1)	
	(1, 1)	
	(1, 1)	
	(2, 1)	
	(2, 1)	
	(2, 1)	
	(2, 1)	

Flight Number	Treatment Combination	Flight Length
	(1, 2)	
	(1, 2)	
	(1, 2)	
	(1, 2)	
	(2, 2)	
	(2, 2)	
	(2, 2)	
	(2, 2)	

Table 10.3 Flight Schedule by Flight Number

Flight Number	Treatment Combination	Flight Length
1	(,)	
2	(,)	
3	(,)	
4	(,)	
5	(,)	
6	(,)	
7	(,)	
8	(,)	

Flight Number	Treatment Combination	Flight Length
9	(,)	
10	(,)	
11	(,)	
12	(,)	
13	(,)	
14	(,)	
15	(,)	
16	(,)	

After a short delay, the random numbers will appear in columns C1 and C2 of the worksheet.

The first row of random numbers gives the treatment combination for the first flight. The level of factor A is in C1 and the level of factor B is in C2. After assigning a treatment combination to the first flight, go to the next row of random numbers and assign a treatment combination to the second flight, and so on. Each time you assign a treatment combination to a flight, write the flight number next to the treatment combination in Table 10.2 and the treatment combination next to the flight number in Table 10.3. *Make sure you do not assign more than four flights to any treatment combination.* For example, if treatment combination (1, 1) already has four flights and the next row of random numbers is (1, 1), skip that row and go to the next row.

There is a small chance that you will run out of random numbers before you have assigned treatment combinations to all of the flights. If this happens, repeat the steps for generating random numbers. After all flights have been assigned a treatment combination, quit Minitab by selecting **Quit** under the **File** menu.

STEP 3: COLLECTING THE DATA

We are now ready to make the flights. For each flight, *make sure the plane is configured according to the proper treatment combination.* After checking that the path is clear, launch the plane. Use a tape measure to measure the shortest distance in inches from the starting point to the plane. Record the distance in the rows of Tables 10.2 and 10.3 corresponding to the flight number.

STEP 4: DATA ANALYSIS

We are now ready to enter our flight data. Insert a diskette and launch Minitab. In the new worksheet, each row will correspond to a flight. Use column **C1** to hold the level (1 or 2) of factor A used on that flight, **C2** for the level of factor B, and **C3** for the length of the flight. Name the columns using the variable name **A** for **C1**, **B** for **C2**, and **Length** for **C3**. Figure 10.2 illustrates the Untitled worksheet at this point.

Enter the data from Table 10.2 into the worksheet. After entering the data for all flights, go through the worksheet and make sure all of the data has been entered correctly. The computer cannot give a correct analysis if you enter incorrect data. Then

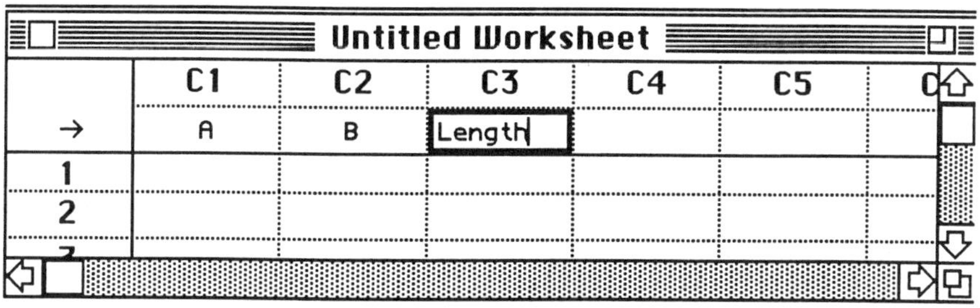

IMPROVING PRODUCT PERFORMANCE WITH PLANNED EXPERIMENTS   173

	C1	C2	C3	C4	C5
→	A	B	Length		
1					
2					

Figure 10.2 Untitled Worksheet

save the worksheet by selecting **Save Worksheet As** under the **File** menu. Name the worksheet **Flight**.

We will want to analyze our data separately for each treatment combination. Minitab will allow us to do separate analyses for subsets of our data where the subsets are formed according to the value in a particular column. As the data on treatment combination is stored in two columns, it will be necessary to create a new column that contains the treatment combination data. We can do this by representing the treatment combination by a two-digit number where the level of factor A is the first (tens) digit and the level of factor B is the second (units) digit. Thus, the new variable will be ten times the entry in C1 plus the entry in C2. We will create the variable in the C4 column. To create the variable:

1. Under the **Calc** menu, click and hold on **Functions and Statistics** and then select **General Expressions** from the submenu. A General Expressions dialog box similar to Figure 10.3 will appear.

2. Click in the box next to **New/modified variable** and type **C4**.

3. Click in the box under **Expression** and type **C1*10+C2**.

4. Click **OK**.

After a short delay, the values of the new variable will appear in column C4 of the worksheet. Name the new variable **Combo**.

We are now ready to analyze our data. Let's first look at separate dotplots for each treatment combination.

1. Under the **Graph** menu, select **Dotplot**. A Dotplot dialog box similar to Figure 10.4 will appear.

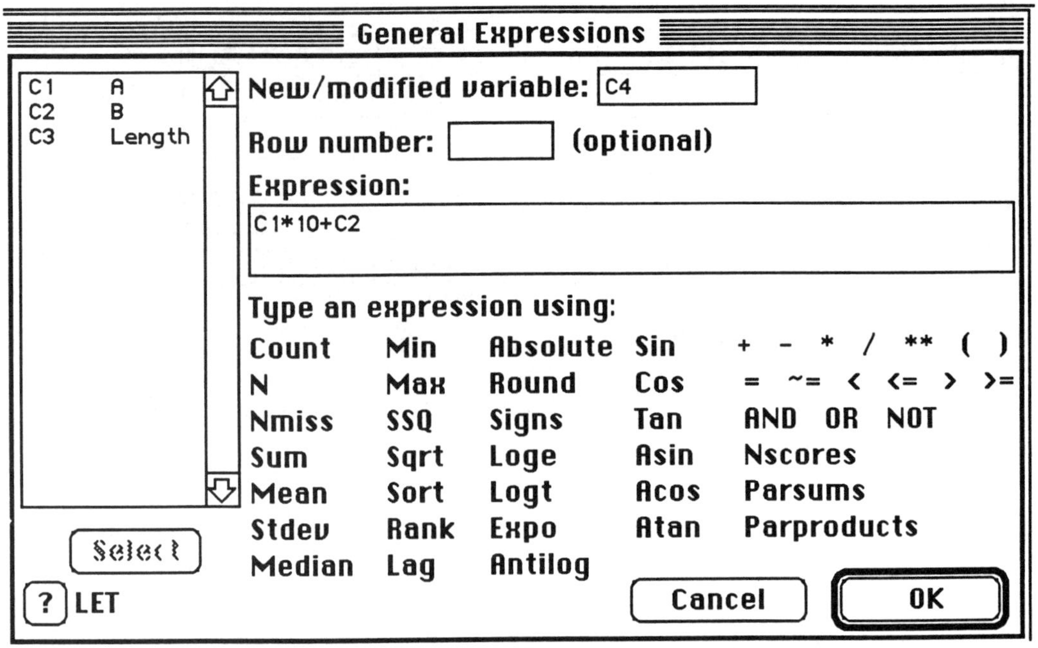

Figure 10.3 General Expressions Dialog Box

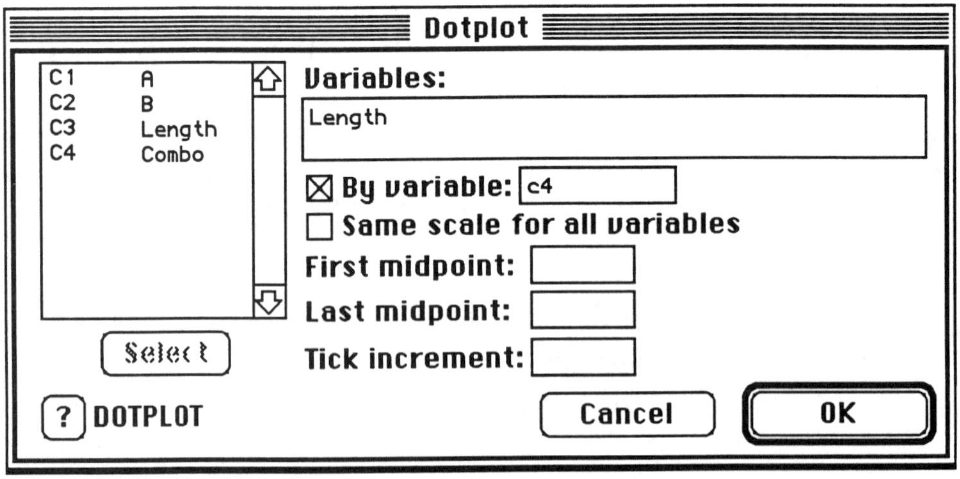

Figure 10.4 Dotplot Dialog Box

2. Click in the box under **Variables** and type **C3** or **Length**.

3. Click in the box to the left of **By variable**.

4. Click in the box to the right of **By variable** and type **C4** or **Combo**.

5. Click **OK**.

Separate dotplots for the variable Length corresponding to each treatment combination will appear on screen. Recall that the treatment combinations are represented by the variable C4. Each flight is represented by a dot. Based on the dotplots, do you think the levels of factors A and B have an effect on the length of flights?

To return to the worksheet, select **Data** under the **Window** menu. Let's now compute descriptive statistics for the sample from each treatment combination.

1. Under the **Stat** menu, click and hold on **Basic Statistics** and then select **Descriptive Statistics** from the submenu. A Descriptive Statistics dialog box similar to Figure 10.5 will appear.

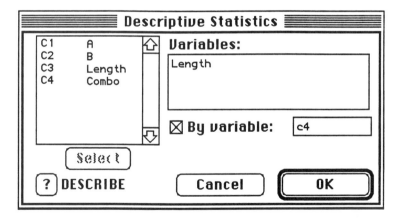

Figure 10.5 Descriptive Statistics Dialog Box

2. Click in the box under **Variables** and type **C3** or **Length**.

3. Click in the box to the left of **By variable.**

4. Click in the box to the right of **By variable** and type **C4** or **Combo**.

Table 10.4 Treatment Combination Means for Flight Distance in Inches

	Factor B Level 1	Factor B Level 2
Factor A Level 1		
Factor A Level 2		

5. Click **OK**.

A set of descriptive statistics will appear in the Session window for each treatment combination.

It is convenient to display the sample means for the four treatment combinations in a table with two rows and two columns. The first row represents level 1 of factor A, and the second row represents level 2 of factor A. The first column represents level 1 of factor B, and the second column represents level 2 of factor B. Write the sample means for the four treatment combinations in your experiment in Table 10.4.

Does it appear that one of the levels of factor A produces longer flights on average? If so, which level yields longer flights? Can you give a physical explanation for this?

Does it appear that one of the levels of factor B produces longer flights on average? If so, which level yields longer flights? Can you give a physical explanation for this?

The last part of our analysis is a graphical investigation of **interaction**. Suppose we consider only level 1 of factor A and plot the mean flight length for each level of factor B. An example graph is given in Figure 10.6. From this graph it is easy to see that for level 1 of factor A, the level of factor B influences the flight distance. The plane flies twice as far with level 2 of factor B.

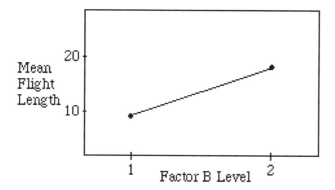

Figure 10.6 Plot of Mean Flight Length with Factor A, Level 1, and Two Levels of Factor B for Example Data

Now envision the same graph with the means plotted for the other level of factor A. Such a graph would have two line segments, one for each level of factor A. Two example graphs are given in Figure 10.7. In the left graph, the effect of the levels of factor B on the flight length for both levels of factor A is similar—level 1 of factor B resulted in shorter flight distances for both levels of factor A. The parallel appearance of these two lines, indicating that factor B has a similar effect on both levels of factor A, is an example of **no interaction**.

The right graph of Figure 10.7 shows a much different scenario. Note that there seems to be an opposite effect of the levels of factor B for the two levels of factor A! This is an example of interaction. It is important to note that the two lines connecting the means do not have to cross to conclude interaction. We look for a highly nonparallel appearance of the lines in determining whether there is interaction.

The presence of interaction can lead to incorrect conclusions if the researcher is not careful. Consider the right graph of Figure 10.7 again. Suppose we averaged the means for the two levels of factor A and plotted them versus the level of factor B. Can you see that this plot would be a rather flat line (both means around 15), suggesting that factor B does not affect flight length? However, this is not the case at all. The effect of the level of factor B on flight length very much depends on the level of factor A that is being used.

178   SESSION TEN

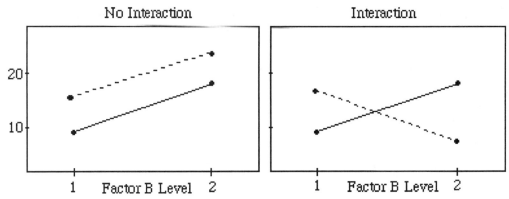

Solid line indicates factor A, level 1; dashed line indicates factor A, level 2.

Figure 10.7 Plots Depicting No Interaction and Interaction of Factors

To produce a graph similar to Figure 10.7, we will create a new Minitab worksheet by selecting **New Worksheet** under the **File** menu. This data set will contain information on the sample mean flight lengths for the four treatment combinations. Use column **C1** to hold the factor A level, **C2** for the factor B level, and **C3** for the sample mean flight length. Name the columns using the variable name **A** for **C1**, **B** for **C2**, and **Means** for **C3**. Enter the factor levels and the appropriate sample means from Table 10.4 in columns C1, C2, and C3. Double-check your entries. Figure 10.8 illustrates the Untitled worksheet for some example data.

	C1	C2	C3
	A	B	Means
1	1	1	8.93
2	1	2	17.65
3	2	1	14.63
4	2	2	21.98
5			
6			

Figure 10.8 Untitled Worksheet

To produce a plot investigating interaction:

1. Under the **Graph** menu, select **Scatter Plot**. A Scatter Plot dialog box similar to Figure 10.9 will appear.

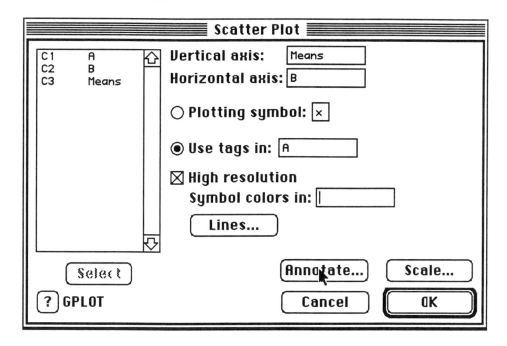

Figure 10.9 Scatter Plot Dialog Box

2. Click in the box to the right of **Vertical axis** and then double-click on **Means**.

3. Click in the box to the right of **Horizontal axis** and then double-click on **B**.

4. To distinguish between the two brands, click on the circle next to **Use tags in**, click in the box to the right of **Use tags in**, and double-click on **A**.

5. Click **Annotate**. An Annotate Scatter Plot dialog box similar to Figure 10.10 will appear.

6. Type entries for the title, footnote, and axis labels in the appropriate boxes.

7. Click **OK** to close the Annotate Scatter Plot dialog box.

8. Click **OK** to end the Scatter Plot command.

The plotting symbol *A* marks the mean for level 1 of factor A and the plotting symbol *B* marks the mean for level 2 of factor A.

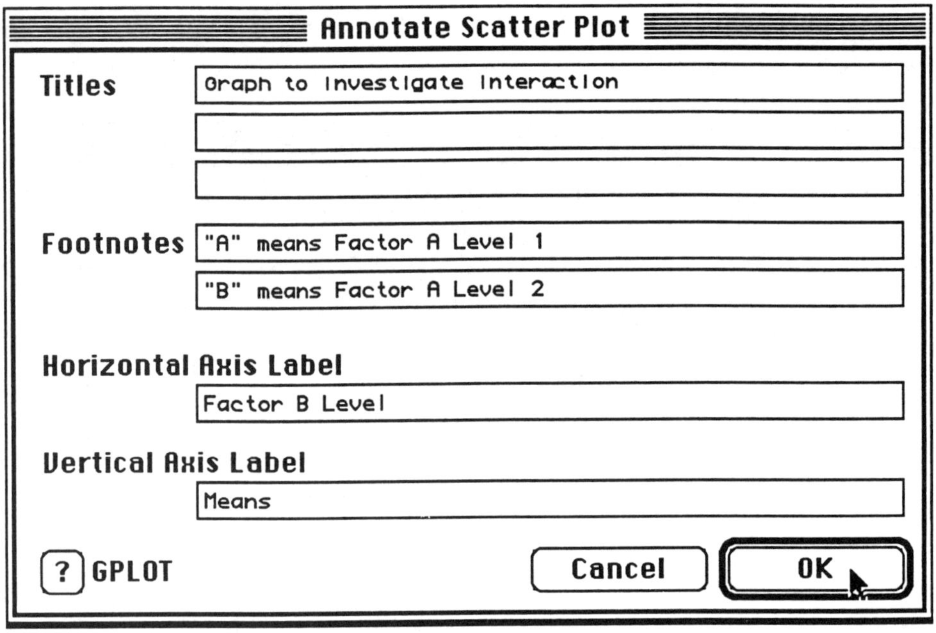

```
╔═══════════════ Annotate Scatter Plot ═══════════════╗
║                                                      ║
║  Titles    ┌──────────────────────────────────────┐ ║
║            │ Graph to Investigate Interaction     │ ║
║            ├──────────────────────────────────────┤ ║
║            │                                      │ ║
║            ├──────────────────────────────────────┤ ║
║            │                                      │ ║
║            └──────────────────────────────────────┘ ║
║                                                      ║
║  Footnotes ┌──────────────────────────────────────┐ ║
║            │ "A" means Factor A Level 1           │ ║
║            ├──────────────────────────────────────┤ ║
║            │ "B" means Factor A Level 2           │ ║
║            └──────────────────────────────────────┘ ║
║                                                      ║
║  Horizontal Axis Label                               ║
║            ┌──────────────────────────────────────┐ ║
║            │ Factor B Level                       │ ║
║            └──────────────────────────────────────┘ ║
║                                                      ║
║  Vertical Axis Label                                 ║
║            ┌──────────────────────────────────────┐ ║
║            │ Means                                │ ║
║            └──────────────────────────────────────┘ ║
║                                                      ║
║  [?] GPLOT              [ Cancel ]     [   OK   ]   ║
╚══════════════════════════════════════════════════════╝
```

Figure 10.10 Annotate Scatter Plot Dialog Box

Print your graph by selecting **Print Window** under the **File** menu. You will need to use a straightedge to connect the means. Make sure you connect *A* to *A* and *B* to *B*, resulting in two lines on your graph. Take a look at your graph. Do you think there is interaction between the two factors?

This concludes our calculations on the data. Make sure you have all the output you need, and then quit Minitab by selecting **Quit** from the **File** menu.

Products are generally improved through a series of experiments. If you found large differences in the effects of the levels of a current factor, that factor is important and you may wish to use other levels of the same factor in future experiments. If you found only small differences in the effects of levels of a current factor, it may not be important and you may wish to experiment with other factors.

Suppose another two-factor experiment with 16 flights is possible. Based on what you have learned in this experiment, what factors and levels would you use?

PARTING GLANCES

Many factors affect the flight distance of a model airplane. We used a planned experiment to study the effects of two of these factors. Some other factors were controlled by holding them constant for all flights. We randomized the order of the flights so no combination of the two factors under study received an unfair advantage from the other factors that we could not control.

We had two levels of two factors with four observations per treatment combination. We could have used other designs for our 16 flights. For example, we could have used two levels of three factors with two observations per treatment combination or two levels of four factors with one observation per treatment combination.

We used descriptive statistics to analyze the results of the experiment. While these give us much useful information, they are not entirely satisfactory. They do not give us the probability that we would have observed differences as large as we did, if in fact the levels of the factors have the same effect on mean flight distance. The area of statistics known as **analysis of variance**, which is beyond the scope of this discussion, deals with these probabilities.

Planned experiments are used throughout industry to improve the quality of products that you use every day. These experiments often involve more than two factors. Consider the bread you eat. A commercial bakery has a large number of potential factors to consider. These include the type of oven, baking time, baking temperature, type of flour, amount of liquid, and so on. Each of these factors may affect important quality characteristics such as taste and shelf life of the bread. By understanding how these factors affect the quality of the bread, the bakery can produce a better product.

An advantage of experiments that involve more than one factor is that we can collect information on all of the factors in a single experiment using the same number of observations required to evaluate a single factor. This leads to major savings in the amount of time and money required for experimentation. Also, a series of one-factor experiments provides no information regarding how factors interact.

EXTENDED WRITING ASSIGNMENT

Refer to Appendix 1, "Technical Report Writing," and Appendix 2, "Technical Report Writing Checklist," for guidance on format and style for your report.

Mr. G. B. "Big" Shot, manager of the Product Design and Development Division, directs you to provide a report summarizing the experiment. The report should include:

1. A summary of the experiment you performed

2. A description of the differences in effects or lack of differences that you found and whether or not the factors interact

3. A recommendation based on current information of the treatment combination that should be used to maximize mean flight distance

4. A recommendation proposing further experimentation on this product

Include graphics where you think they are helpful.

Name   Section   Session 10

SHORT ANSWER WRITING ASSIGNMENT

All answers should be complete sentences. Include copies of Table 10.4 and your interaction graph with this assignment.

1. What were the levels of factor A and factor B in your experiment?

2. Why did we attempt to keep the launching force constant for all flights?

3. Why did we randomize the order of the 16 flights?

4. Based on your results, do you think that the level of factor A has an effect on mean flight distance? If so, what level of this factor do you recommend?

5. Based on your results, do you think that the level of factor B has an effect on mean flight distance? If so, what level of this factor do you recommend?

6. Based on your interaction graph, do you think that factors A and B interact?

7. What experiment would you suggest to further improve the mean flight distance?

8. Propose two factors to be used in an experiment involving the production of a cola drink.

SESSION ELEVEN

Breaking Strength of Facial Tissue

INTRODUCTION

A common quality characteristic of manufactured products is their strength. Strength is important in such varied items as fibers, paper, cardboard, concrete, plastics, and metals. One method of measuring an object's strength is to put it under continually increasing stress until it breaks. The amount of stress on the object when it breaks is called its **breaking strength**.

In this experiment we will measure the breaking strength of one-ply facial tissues. The stress will be the force of fishing weights placed on the tissue, and the measurement will be the number of weights required to tear the tissue enough that at least one weight falls through the tear.

STATISTICAL CONCEPTS

Stem-and-leaf plot, normality assumption, one-sample t-test and confidence interval, nonparametric sign confidence interval.

MATERIALS NEEDED

For each team, a 7-inch embroidery hoop, 25 one-ply facial tissues with a minimum dimension of at least 8 inches, 3 12-ounce soft drink cans, and 20 1-ounce egg-shaped fishing sinkers.

THE SETTING

You are a quality engineer for the It's Your Face tissue company. A production operator has given you a sample of 20 tissues collected from the manufacturing process. You are to collect and analyze breaking-strength data for the sample of tissues. There is particular interest in the mean breaking strength. The target value for mean breaking strength is 7 weights. If the mean is too low, the tissues tear too easily. If it is too high, then other quality characteristics such as softness tend to deteriorate.

BACKGROUND

Previous experiments suggest that there may be a large amount of variability in the breaking-strength measurements. Some of this variation is due to differences among the tissues. These differences include variation in the number and types of flaws in the tissue fibers and the alignment and sizes of the fibers within the tissue.

Other variation is due to the measurement process. A tissue is much more likely to tear if the weights are dropped on the tissue than if they are gently placed. The tissue is more likely to tear if all of the weights are placed at the same location than if they are spread evenly across the same tissue. Also, the rate at which the weights are put on the tissue will have an effect on the measurement. For example, if a tissue is capable of withstanding a stress of 4 weights for exactly 7 seconds before it breaks, then the measurement will be 4 if more than 7 seconds pass between the placement of the 4th and 5th weights. The measurement will be greater than 4 if less than 7 seconds pass between the placement of these weights.

To get a clear understanding of the strength distribution for a sample of tissues, we want to eliminate variation due to the measurement process. This is done by defining a precise plan for taking the measurements and carefully following it. Without such a plan, it is difficult to compare measurements taken on different tissues, particularly if the measurements are taken at different times by different people. The plan for taking measurements is called a measurement **protocol**.

THE EXPERIMENT

STEP 1: THE BASIC MEASUREMENT

The breaking strength of a tissue will be measured while the tissue is being held in place by an embroidery hoop. An embroidery hoop, which is illustrated in Figure 11.1, consists of inner and outer hoops that are held together by a clamp on the outer hoop.

The clamp is tightened by turning the screw clockwise and loosened by turning the screw counterclockwise.

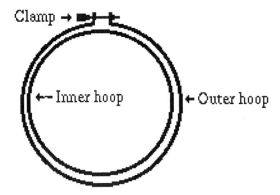

Figure 11.1 Embroidery Hoop

The first step of the measurement process is to clamp the tissue in the embroidery hoop. Before placing the tissue in the hoop, *make sure it is a one-ply tissue.* If it is two-ply, separate it and use only one ply. Separate the outer and inner hoops. Completely cover the inner hoop with the tissue and place the inner hoop inside the outer hoop. Pull *gently* on each corner of the tissue to eliminate wrinkles. Tighten the outer hoop, making sure the tissue does not wrinkle. If the tissue wrinkles, loosen the screw, pull on the corners, and tighten the screw. If the tissue tears while it is being placed in the hoop, discard it and use another tissue. When the tissue is securely clamped in the hoop, place the hoop on three soda cans as shown in Figure 11.2.

Figure 11.2 Embroidery Hoop on Soda Cans

The second step of the measurement process is placing the weights on the tissue. *Before you place the weights on the tissue,* make sure you completely understand how they are to be placed. The first weight is to be gently placed in the center of the hoop. Continue to gently place additional weights on the tissue, such that each additional weight is placed as close to the center as possible without stacking the weights or covering the frame of the hoop. Allow approximately 3 seconds to pass before the placement of each additional weight. This is approximately the amount of time required to say "one thousand one, one thousand two, one thousand three." Figure 11.3 illustrates a hoop with eight weights on it. If an additional weight cannot be added without stacking or covering the frame of the hoop, then start a second layer of weights beginning as close to the center as possible.

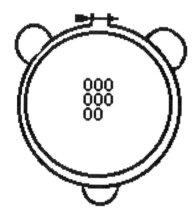

Figure 11.3 Arrangement of Eight Weights on Hoop

The measurement of interest is the number of weights on the tissue when it tears enough to allow at least one weight to fall through the tear. If the tissue tears as a weight is being put on it, include that weight in the count. To make sure your team understands the measurement protocol, measure the breaking strength of three practice tissues. These three tissues are for practice only and will not be used in the data analysis.

If, after making the practice measurements, you are comfortable with the measurement protocol, move on to Step 2. Otherwise, clarify any uncertainty with your team members and your teacher.

Table 11.1 Breaking Strengths of Tissues

Tissue	1	2	3	4	5	6	7	8	9	10
Strength										

Tissue	11	12	13	14	15	16	17	18	19	20
Strength										

STEP 2: COLLECTION OF THE DATA

Use the measurement protocol to measure the breaking strength for the sample of 20 tissues that you were given. Have the same team member place the weights on all tissues. This will eliminate unnecessary person-to-person variability in the measurement process. Be sure to follow the protocol as closely as possible. Record the measurements in Table 11.1.

STEP 3: DATA ANALYSIS

Insert a diskette and launch Minitab. Enter your breaking-strength data in column **C1** using the variable name **Strength**. Figure 11.4 illustrates the Untitled worksheet with some example data. Double-check the entered data to make sure it is correct. After you are sure it is correct, save your data on your diskette by selecting **Save Worksheet As** under the **File** menu. Name your data **Tissues**.

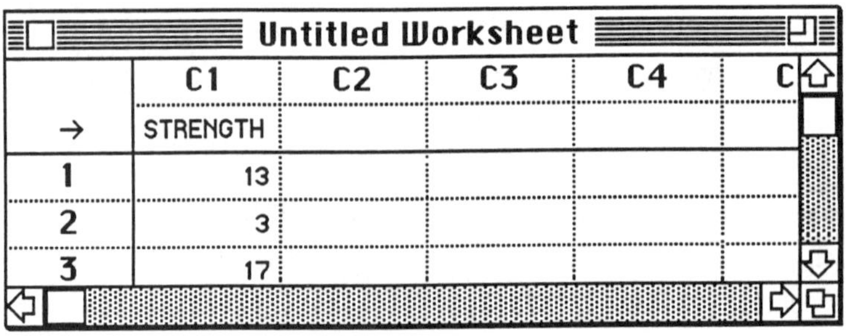

	C1	C2	C3	C4
→	STRENGTH			
1	13			
2	3			
3	17			

Figure 11.4 Untitled Worksheet

Let's begin our data analysis by looking at some descriptive statistics. To do this:

1. Under the **Stat** menu, click and hold on **Basic Statistics** and then select **Descriptive Statistics** from the submenu. A Descriptive Statistics dialog box similar to Figure 11.5 will appear.

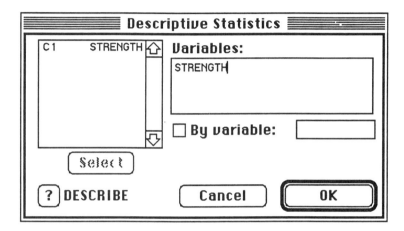

Figure 11.5 Descriptive Statistics Dialog Box

2. Click in the box under **Variables** and double-click on **Strength**.

3. Click **OK**.

Descriptive statistics will appear in the Session window. Look at the results. Are they what you expected based on your data? If not, you may have entered the data incorrectly. If you want to check your data again, select **Data** under the **Window** menu. If you have errors, correct them, save the changes by selecting **Save Worksheet** under the **File** menu, and produce a new set of descriptive statistics. After you are sure your descriptive statistics are correct, complete Table 11.2.

Let's now learn about the shape of the distribution by producing a stem-and-leaf diagram. To do this:

1. Under the **Graph** menu, select **Stem-and-Leaf**. A Stem-and-Leaf dialog box similar to Figure 11.6 will appear.

Table 11.2 Descriptive Statistics for Breaking Strength

Sample size	
Sample mean	
Sample median	
Sample standard deviation	
Sample minimum	
Sample maximum	

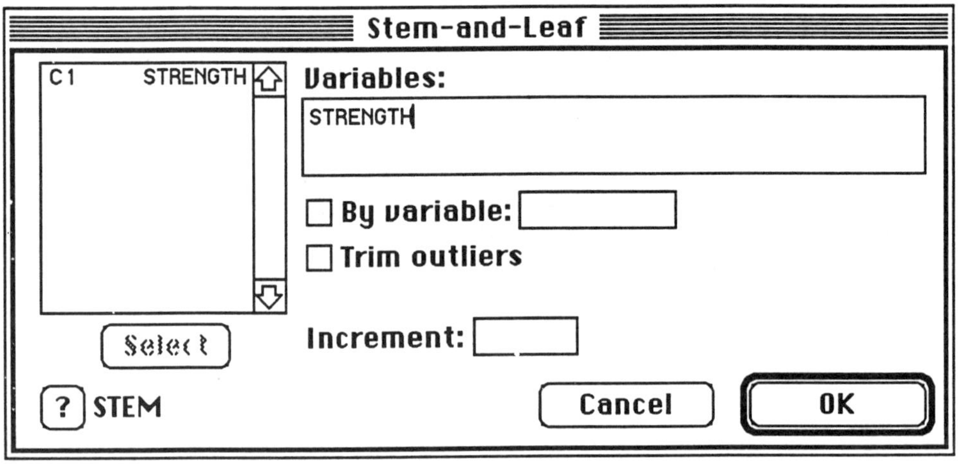

Figure 11.6 Stem-and-Leaf Dialog Box

2. Click in the box under **Variables** and double-click on **Strength**.

3. Click **OK**.

The stem-and-leaf diagram for the variable Strength will appear in the Session window. You could have also produced a histogram, dotplot, or boxplot using the Graph menu.

The stem-and-leaf diagram shows information that is not given in Table 11.2. It allows us to look at the shape of the distribution of breaking strengths. How would you describe the shape of the distribution? Does the distribution appear to be approximately normal?

Now, get an alternative view of this data by selecting **Histogram** under the **Graph** menu. Follow items 2 and 3 from the instructions for producing the stem-and-leaf diagram. The histogram gives you a different view of the data. Have your opinions about the shape of the distribution of breaking strengths changed?

The It's Your Face tissue company strives for a population mean breaking strength of seven weights. Let's use Minitab to find a 95 percent one-sample *t* confidence interval for the mean breaking strength for the population from which our sample was taken. Denote this population mean by µ. Remember that when working with small samples, the one-sample *t* confidence interval is based on the assumption that the individual measurements are a random sample from a normal distribution. If the stem-and-leaf diagram suggests a marked departure from normality, then the true degree of confidence may be different from 95 percent.

To compute the one-sample *t* confidence interval in Minitab:

1. Under the **Stat** menu, click and hold on **Basic Statistics** and then select **1-Sample t** from the submenu. A 1-Sample t dialog box similar to Figure 11.7 will appear.

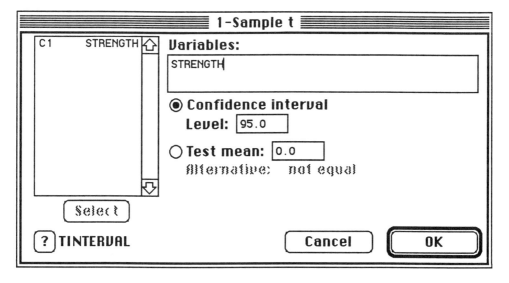

Figure 11.7 1-Sample t Dialog Box

2. Click in the box under **Variables** and double-click on **Strength**.

3. Verify that the button immediately to the left of **Confidence interval** is selected and that the box to the right of **Level** contains 95.0. If not, click the button, click in the box, and type **95.0**.

4. Click **OK**.

The one-sample *t* confidence interval will appear in the Session window. Write the resulting confidence interval on the line below:

95% confidence interval for population mean: (,   ).

If, for example, the interval is (5.16, 6.75), this would suggest that we are 95 percent confident that the true value of the population mean is between 5.16 and 6.75.

There is a relationship between the confidence interval and hypothesis tests with two-sided alternatives. For this experiment the tissue company wants the population mean to be 7.0. If the value 7.0 is within the 95 percent confidence interval, then the one-sample t-test would not reject the null hypothesis of $\mu = 7.0$ versus the two-sided alternative $\mu \neq 7.0$ with a 5 percent significance level. If 7.0 is not within the confidence interval, then the null hypothesis $\mu = 7.0$ would be rejected in favor of the two-sided alternative $\mu \neq 7.0$. If the entire interval is above 7.0, then we can conclude not only that $\mu \neq 7.0$, but also that $\mu > 7.0$. If the entire interval is below 7.0, then we can conclude not only that $\mu \neq 7.0$, but also that $\mu < 7.0$. Note that our example confidence interval (5.16, 6.75) is entirely below 7.0.

To illustrate this relationship and to show how one-sample t-tests are done in Minitab:

1. Under the **Stat** menu, click and hold on **Basic Statistics** and then select **1-Sample t** from the submenu. A 1-Sample t dialog box similar to Figure 11.8 will appear.

2. Click in the box under **Variables** and double-click on **Strength**.

3. Click the button immediately to the left of **Test mean**, and then click in the box to the right of **Test mean** and type **7.0**.

4. Verify that the alternative listed below **Test mean** is **not equal**. If it is not, drag down on the arrow in the Alternative box and select **not equal**.

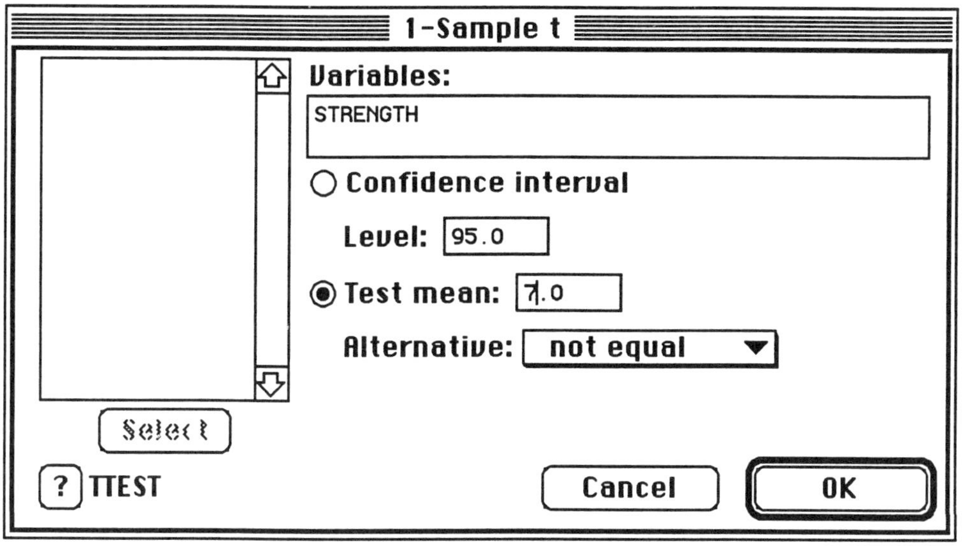

BREAKING STRENGTH OF FACIAL TISSUE

```
┌─────────────────────────────────────────────┐
│              ≡ 1-Sample t ≡                 │
│  ┌──────┐ ⇧  Variables:                     │
│  │      │    ┌──────────────────────────┐   │
│  │      │    │ STRENGTH                 │   │
│  │      │    └──────────────────────────┘   │
│  │      │    ○ Confidence interval          │
│  │      │       Level: [95.0]               │
│  │      │ ⇩  ● Test mean: [7.0]             │
│  └──────┘    Alternative: [ not equal  ▼]   │
│   [Select]                                  │
│  [?] TTEST              [Cancel]   [  OK  ] │
└─────────────────────────────────────────────┘
```

Figure 11.8 Second 1-Sample t Dialog Box

5. Click **OK**.

The hypotheses, the value of the one-sample t statistic, and the p-value of the test will appear in the Session window. If the p-value is less than or equal to 0.05, one would reject the null hypothesis that $\mu = 7$ versus a two-sided alternative at the 5 percent significance level. If the p-value is greater than 0.05, you would not reject the null hypothesis. Write the value of the t statistic, the p-value, and your decision on the line below.

$t = $ _____ , p-value = _____ , decision: _____

Is your decision consistent with the results of the confidence interval (i.e., reject the null hypothesis if 7 is not within the interval)?

What does the result of your hypothesis test tell us about the population of facial tissues?

If we had wanted to use a one-sided alternative, we would have dragged down on the arrow in the Alternative box and clicked on the appropriate one-sided alternative.

The one-sample t confidence interval and hypothesis test are based on the assumption that the individual measurements follow a normal distribution. There are a type of statistical methods, known as **nonparametric statistics**, that do not require the assumption of normally distributed data. The one-sample sign procedure that is available in Minitab yields a confidence interval for the population median.

The one-sample sign confidence interval is based on the fact that for continuous distributions, the number of sample observations that are less than or equal to the population median has a binomial distribution. The binomial parameters are n = the sample size and $p = 0.5$. Remember that the population median is the midpoint of the ordered population values. This may be different from the population mean, which is the average value in the population. Due to the discreteness of the binomial distribution, it is generally not possible to achieve a confidence level of exactly 95 percent. For our example with a sample size of 20, the interval has a 95.86 percent degree of confidence.

To calculate the intervals by hand, one would order the observations from smallest to largest. The endpoints of the interval are specific observations in the ordered data set. With a sample of size 20 and a desired degree of confidence of 95.86 percent, the endpoints are the 6th and 15th observations in the ordered data set.

To have Minitab calculate the one-sample sign confidence interval for the population median having degree of confidence as close as possible to 95 percent:

1. Under the **Stat** menu, click and hold on **Nonparametrics** and then select **1-Sample Sign** in the submenu. A 1-Sample Sign dialog box similar to Figure 11.9 will appear.

2. Click in the box under **Variables** and double-click on **Strength**.

3. Verify that the button immediately to the left of **Confidence interval** is selected and that the box to the right of **level** contains **95.0**. If not, click the button, click in the box, and type **95.0**.

4. Click **OK**.

The one-sample sign confidence intervals having the largest possible degree of confidence smaller than 95 percent and the smallest possible degree of confidence

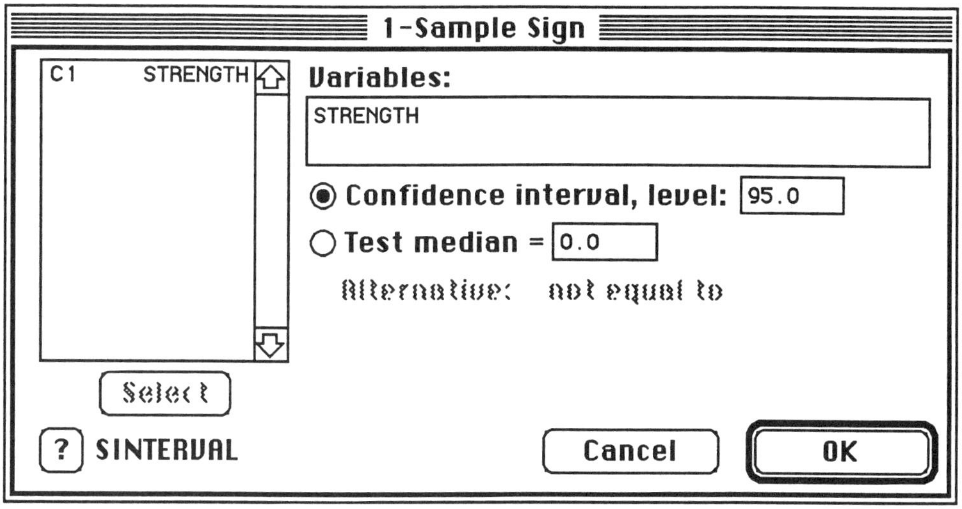

Figure 11.9 1-Sample Sign Dialog Box

larger than 95 percent are given in the Session window. Minitab also interpolates between the two intervals to approximate a 95 percent interval. Write the resulting 95.86 percent confidence interval for the population median on the line below:

95.86% confidence interval for population median: (,   ).

Print your Session window by selecting **Print Window** under the **File** menu. After you get your output, quit Minitab by selecting **Quit** under the **File** menu.

PARTING GLANCES

Our breaking-strength measurements were discrete in that they were measured in terms of the number of weights required to tear the tissue. In many industrial applications, an expensive device designed to produce a continually increasing amount of force on the object would be used. With this type of test equipment, the breaking-strength measurement would be continuous.

We used two methods, the t and the sign, to produce confidence intervals for the center of the population. They differed in that the t interval is an interval for the population mean and the sign interval is an interval for the population median. For small samples, the t interval requires the assumption that the data came from a normal distribution. Minitab also allows for the computation of one-sample Wilcoxon confidence intervals. This procedure is based on the assumption that the data comes from a symmetric, continuous, but not necessarily normal, distribution. The one-sample Wilcoxon procedure is in the same submenu as the one-sample sign procedure.

EXTENDED WRITING ASSIGNMENT

Refer to Appendix 1, "Technical Report Writing," and Appendix 2, "Technical Report Writing Checklist," for guidance on format and style for your report.

Write a report to the production manager, I. Y. Face, Jr., summarizing the experiment. It should include:

1. A description of the experiment including a clear explanation of the measurement protocol for breaking strength

2. A graphic and numerical summary of the results of your experiment

3. A description of the confidence intervals and hypothesis test that were performed and the implications of these results to the It's Your Face tissue company

4. Any recommendations you might have regarding changes in the measurement protocol or for future experimentation

Name _____ Section _____ Session 11

SHORT ANSWER WRITING ASSIGNMENT

All answers should be complete sentences. Include a copy of your stem-and-leaf diagram with this assignment.

1. Explain how the measurement of breaking strength was made. Be specific enough so that someone not present when the experiment was performed will understand.

2. Based on your stem-and-leaf diagram, do you believe it is reasonable to assume that the breaking-strength data comes from a normal distribution?

3. Based on your answer to question 2, do you believe that the assumptions of the one-sample t confidence interval and hypothesis test have been satisfied?

4. Provide the following descriptive statistics for your strength data:

    Sample size _____, sample mean _____,

    sample median _____, sample standard deviation _____.

5. Give the 95 percent one-sample t confidence interval for the population mean breaking strength and explain the meaning of these results for the It's Your Face tissue company.

6. Give the value of the t statistic for testing the null hypothesis that the population mean breaking strength equals 7 versus a two-sided alternative and the results of your test using a 5 percent significance level.

7. The nonparametric confidence interval does not require us to assume that breaking strength follows a normal distribution. Is this an advantage over the t confidence interval?

8. The one-sample t confidence interval estimates the population mean. What population parameter is estimated by the one-sample sign confidence interval?

SESSION TWELVE

Conclusion of Plant-Growth Experiment

INTRODUCTION

A good experiment should be carefully planned from the very beginning, long before any data is obtained. Careful attention should be given to the choice of the factors and the levels of each factor. Deciding on the experimental units and how they are assigned to the treatment combinations is also very important. Well-designed experiments can be found in many areas of science, business, and industry. In psychology, experiments are used to determine whether a new IQ test is more predictive of a person's academic achievement than the standard one. A biologist might plan an experiment to determine whether water temperature or type of food significantly affects the growth of a particular species of fish.

STATISTICAL CONCEPTS

Two-factor design, treatment combinations, multiple scatter plots.

MATERIALS NEEDED

Completed Table 4.4.

THE SETTING

You are a member of a small agricultural research team working for Harvest Veg, a company that grows vegetables organically without the use of pesticides. In Session 4 you began an experiment to study the effect of seed type and amount of water on the growth of a particular type of plant. You have 12 plants, illustrating all combinations of two watering levels and two seed varieties. You are now ready to analyze your data, the weekly plant heights of each plant, and make a report to Harvest Veg President Irene Bean.

BACKGROUND

Harvest Veg normally grows their plants in a prepared medium that is a dark soil rich in nutrients. It has always been thought that this soil consistently produces the

Table 12.1 Treatment Combinations for Plant-Growth Experiment

Symbol	Levels	Detailed Description	General Description
A	(1, 1)		Seed variety 1 and water level 1
B	(1, 2)		Seed variety 1 and water level 2
C	(2, 1)		Seed variety 2 and water level 1
D	(2, 2)		Seed variety 2 and water level 2

largest and healthiest vegetables. Recently, some field technicians noticed variability in the crops. Some plants appeared healthy and large, while others grown in the same field were very small. This is what prompted you to design and carry out this experiment. You would like to find a combination of seed variety and watering amount that produces uniformly large plants.

THE EXPERIMENT (CONTINUED)

STEP 1: DATA ENTRY

It is important to recall what the four treatment combinations are for your experiment. Complete Table 12.1 using the information in Table 4.1. Recall that in the first part of this experiment we used randomization to decide the arrangement of the 12 pots in the window. In Figure 12.1, indicate the arrangement of treatment combinations (A, B, C, and D) for the 12 pots. This information is given in Figure 4.3.

Before the data is entered into a Minitab worksheet, you should first convert the data in Table 4.4 to decimals. Two decimal places should be sufficient. You can use

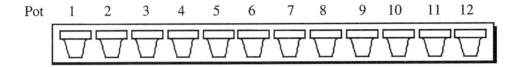

Figure 12.1 Treatment-Combination Assignments to Pots

the Macintosh Calculator or your own, if needed. Make sure you have a weekly measurement for all 12 plants and that all measurements are in centimeters.

We will store our data in a Minitab worksheet with 13 columns, one for each of the 12 plants and one for the week number. Each worksheet row will contain the 12 measured heights for a particular week and the week number: 1, 2, 3, and so on. There will be as many rows as the number of weeks that you collected data. For example, if your Table 4.4 contains data for 7 weeks, your worksheet will have 7 rows.

We will use Minitab variable names that clearly identify which levels of the two factors are represented in a pot by using the following two-character naming procedure for the 12 pots:

Variable name's first character denotes treatment combination: A, B, C, or D

Variable name's second character identifies replicate: 1, 2, or 3

For example, if pot 1 contained the first replicate of seed variety 2 and water level 2 (i.e., treatment combination D), it would be named **D1**.

Choose a Macintosh and launch Minitab. Name the columns **C1–C12** using the appropriate two-character names for the 12 pots and give column **C13** the name **Week**. Figure 12.2 illustrates the variable names in the Minitab worksheet for an example assignment of treatment combinations to pots. Yours will most likely be different.

	C1	C2	C3	C4	•••	C10	C11	C12	C13
→	B1	C1	C2	A1		D3	C3	A3	WEEK
1	3.5	3.00	3.00	1.5	•••	4.25	4.00	3.25	1
2	4.0	3.75	3.25	2.0	•••	5.00	4.25	4.50	2

Figure 12.2 Variable Names for Example Assignment of Treatment Combinations to Pots

Enter the data from Table 4.4 into the worksheet. After the data has been entered, carefully check each data value, and then save the worksheet on your diskette by selecting **Save Worksheet As** under the **File** menu. Name the worksheet **Plant.height**.

STEP 2: DATA ANALYSIS

We will begin our analysis by calculating weekly mean heights for the four treatment combinations $A, B, C,$ and D. Begin by calculating the sample mean for the three plants from treatment combination A. You will need to create a new column of average weekly heights for the three plants **A1**, **A2**, and **A3**. To do this using Minitab:

1. Under the **Calc** menu, click and hold on **Functions and Statistics** and then select **Row Statistics** from the submenu. A Row Statistics dialog box similar to Figure 12.3 will appear.

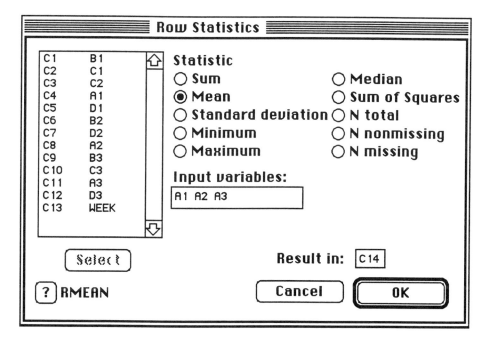

Figure 12.3 Row Statistics Dialog Box

Table 12.2 Average Plant Height in Centimeters at End of Week 3 and End of Experiment

	Week 3		Experiment End	
	Water 1	Water 2	Water 1	Water 2
Seed variety 1				
Seed variety 2				

2. Click in the box under **Input variables** and then double-click on each of **A1**, **A2**, and **A3**.

3. Click in the circle next to **Mean**.

4. Click in the box next to **Result in** and type **C14**.

5. Click **OK**.

A new variable that represents the weekly average heights of the three plants with seed variety 1 and water level 1 will now appear in your worksheet in column C14. Give this variable the name **Amean**.

Now follow the same procedure for the other three treatment combinations. Store these weekly average heights in **C15**, **C16**, and **C17** and give these variables the names **Bmean**, **Cmean**, and **Dmean**, respectively. In computing Cmean make sure you double click on the *variable names* C1, C2, and C3 and not on the *column names* C1, C2, and C3. After you have created these new variables, save the changes to your worksheet by selecting **Save Worksheet** under the **File** menu.

By looking at these four new columns we may discover that one treatment combination produced the tallest plants. Or perhaps what was tallest midway through the experiment did not end up being the best in the end. Use the appropriate entries in these new columns to complete Table 12.2. Look carefully at Table 12.2. Based on these averages, does it appear that one seed variety grows better than the other? Is this true for both of your water levels?

Let's now produce several graphs of the growth curves, which are just plots of plant height versus week. For purposes of comparison, it will be useful to plot more than one curve on the same plot, so the **Multiple Scatter Plot** procedure in Minitab will be used. We shall begin with a time plot of the weekly heights for the three plants for treatment combination *A*. To do this using Minitab:

1. Under the **Graph** menu, select **Multiple Scatter Plot**. A Multiple Scatter Plot dialog box similar to Figure 12.4 will appear.

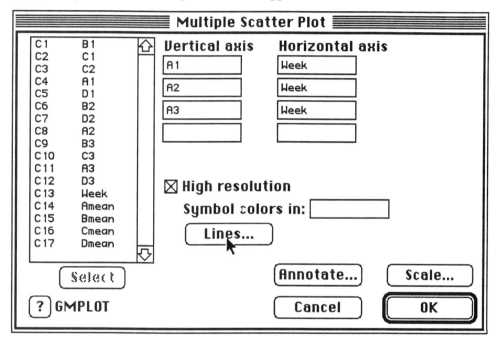

Figure 12.4 Multiple Scatter Plot Dialog Box

2. Click in the first box under **Vertical axis** and type **A1**.
3. Click in the first box under **Horizontal axis** and type **Week**.
4. Click in the second box under **Vertical axis** and type **A2**.
5. Click in the second box under **Horizontal axis** and type **Week**.
6. Click in the third box under **Vertical axis** and type **A3**.
7. Click in the third box under **Horizontal axis** and type **Week**.

8. Click **Lines**. A Lines dialog box similar to Figure 12.5 will appear.

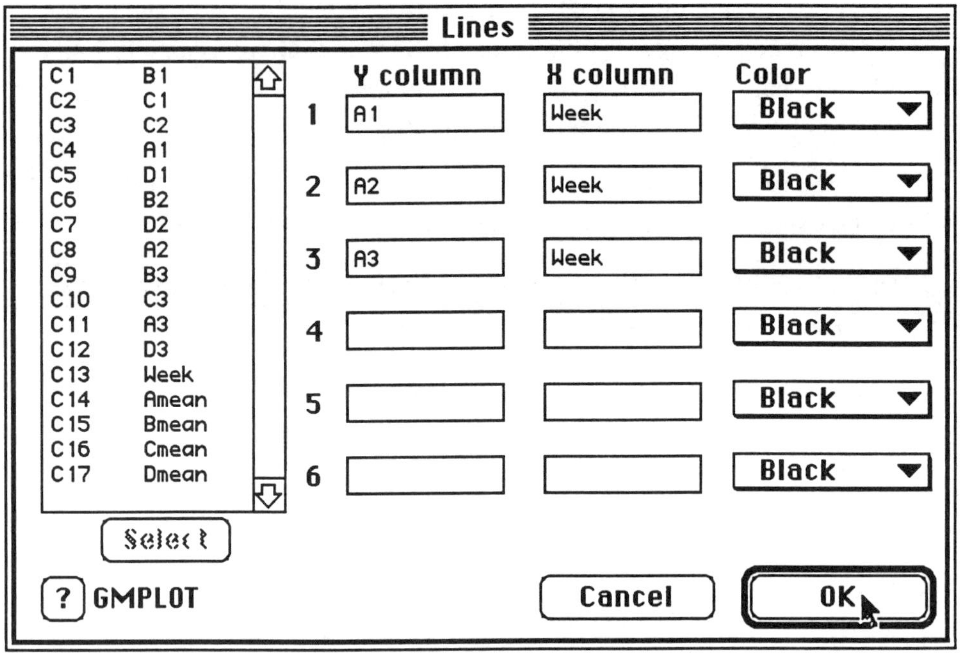

Figure 12.5 Lines Dialog Box

9. Enter the variable names **A1**, **A2**, and **A3** in the first three boxes under **Y column** and enter **Week** in each of the first three boxes under **X column**.

10. Click **OK** to return to the Multiple Scatter Plot dialog box.

11. Click **Annotate**. An Annotate Scatter Plot dialog box similar to Figure 12.6 will appear.

12. Enter an appropriate title, footnote, and axis labels, and click **OK** to return to the Multiple Scatter Plot dialog box.

13. Click **OK**.

Get a "hard copy" of the graph by selecting **Print Window** under the **File** menu. Next, obtain similar graphs, each showing three growth curves, for the other treatment combinations, *B, C,* and *D*. Remember to get printouts of these graphs.

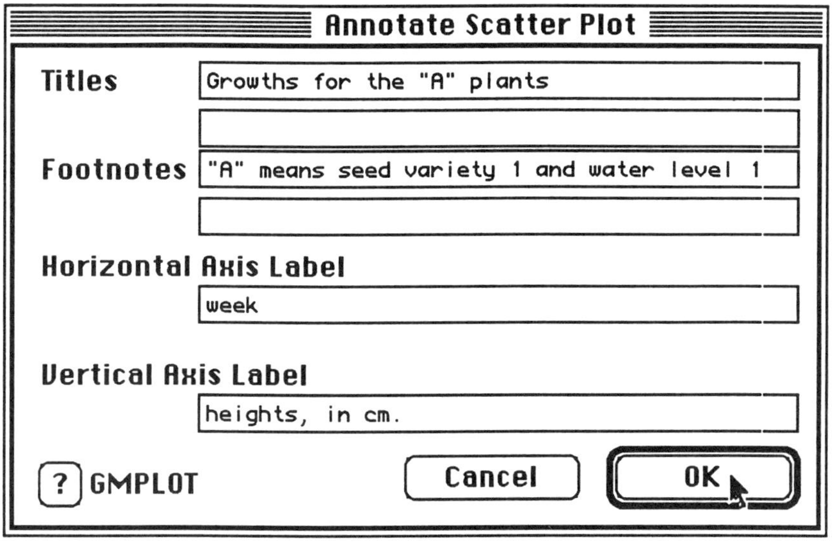

```
┌─────────────────────────────────────────────────────┐
│≡≡≡≡≡≡≡≡≡≡≡≡≡≡≡ Annotate Scatter Plot ≡≡≡≡≡≡≡≡≡≡≡≡≡≡│
│ Titles    ┌─────────────────────────────────────┐   │
│           │ Growths for the "A" plants          │   │
│           ├─────────────────────────────────────┤   │
│           │                                     │   │
│ Footnotes ├─────────────────────────────────────┤   │
│           │ "A" means seed variety 1 and water level 1│
│           ├─────────────────────────────────────┤   │
│           │                                     │   │
│           └─────────────────────────────────────┘   │
│ Horizontal Axis Label                               │
│           ┌─────────────────────────────────────┐   │
│           │ week                                │   │
│           └─────────────────────────────────────┘   │
│ Vertical Axis Label                                 │
│           ┌─────────────────────────────────────┐   │
│           │ heights, in cm.                     │   │
│           └─────────────────────────────────────┘   │
│   ┌───┐                  ┌──────────┐  ┌──────────┐ │
│   │ ? │ GMPLOT           │  Cancel  │  │   OK     │ │
│   └───┘                  └──────────┘  └──────────┘ │
└─────────────────────────────────────────────────────┘
```

Figure 12.6 Annotate Scatter Plot Dialog Box

Examine these four graphs closely. For each graph, the three plants had the same seed variety and received the same amount of water, so they should be fairly similar in their growth. Does there appear to be excessive variability *within* treatment combinations? For each plot, do the three plants consistently get taller each week?

Remember that we are ultimately trying to determine which treatment combination(s) produced the tallest plants. It may be useful to place these four plots side by side, studying them carefully. Did one seed variety grow better than the other? On the other hand, perhaps water really makes a difference but seed variety doesn't.

Four more multiple scatter plots will facilitate comparison and make it easier to answer the questions posed above. For these plots, we use the variables Amean, Bmean, Cmean, and Dmean, the average plant heights for the four treatment combinations.

We first plot Amean versus Week and Bmean versus Week on the same graph. This helps to compare the two water levels when using seed variety 1.

1. Under the **Graph** menu, select **Multiple Scatter Plot**. A Multiple Scatter Plot dialog box similar to Figure 12.7 will appear.

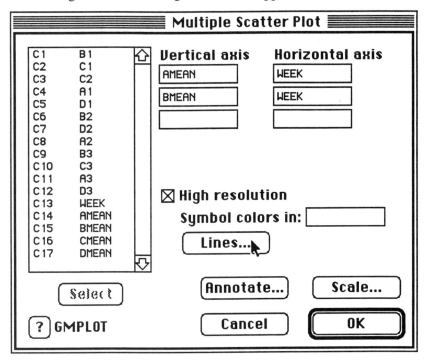

Figure 12.7 Multiple Scatter Plot Dialog Box

2. Enter the variable names **Amean** and **Bmean** in the first two **Vertical axis** boxes, as shown in Figure 12.7.

3. Enter the variable name **Week** in the first two **Horizontal axis** boxes.

4. Click **Lines** and enter these variable names in the **Y column** and **X column** boxes as shown in Figure 12.8.

5. Click **Annotate** and enter an appropriate title, footnote, and axis labels.

6. Click **OK** to return to the Multiple Scatter Plot dialog box.

7. Click **OK**.

CONCLUSION OF PLANT-GROWTH EXPERIMENT   211

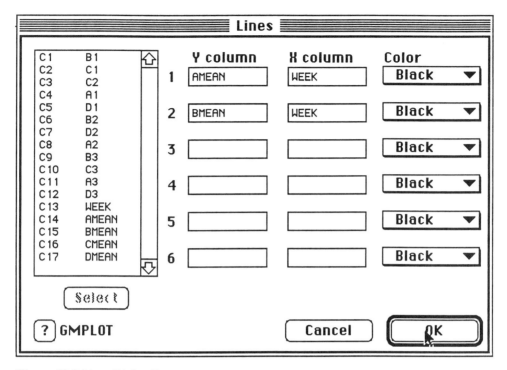

Figure 12.8 Lines Dialog Box

If you are satisfied with your plot, print it by selecting **Print Window** under the **File** menu. Does it appear that the watering scheme had an effect on the plants with seed variety 1?

Now, repeat this procedure with modifications to plot Cmean versus Week and Dmean versus Week on the same graph. This plot will allow you to compare average plant growth for the two water levels when using seed variety 2. Print your plot. Does it appear that the watering scheme affected the plants with seed variety 2?

Plot Amean versus Week and Cmean versus Week on the same graph. This plot will allow you to compare average plant growth for the two seed varieties when using water level 1. Print your plot. Does it appear that the seed variety affected the plants receiving water level 1?

Plot Bmean versus Week and Dmean versus Week on the same graph. This plot will allow you to compare average plant growth for the two seed varieties when using water level 2. Print your plot. Does it appear that the seed variety affected the plants receiving water level 2?

Now, plot the average growth for each of the four treatment combinations on the same graph. Use the procedure described above with Amean versus Week, Bmean versus Week, Cmean versus Week, and Dmean versus Week on the same graph. What patterns do you see in this graph?

This concludes our calculations on the data. Make sure you have all the output you need, and then quit Minitab by selecting **Quit** from the **File** menu.

PARTING GLANCES

Data from a designed experiment was analyzed to investigate the differences in plant growth. By examining what caused the variability in the plant heights, we hope to discover an optimal treatment combination.

If one of your factors did not make a difference, do not think that your experiment was a failure. For example, if your two levels of the water factor produce plants of very similar size, this would imply that the levels chosen for water were not a significant source of variability. This may suggest that in future experiments we should choose two more widely separated water levels.

There are many other important uses of our plant data. Another reason we look at plots is to examine the behavior of the growth curves. Does it appear from our plots that growth appears linear over time, or is the relationship one of curvature? If we were interested in finding an equation that predicted plant height as a function of week, we could use curve-fitting techniques to model the observed growth curves. Advanced regression techniques could also be used to quantify how much faster or taller plants from one treatment combination grew than another.

EXTENDED WRITING ASSIGNMENT

Refer to Appendix 1, "Technical Report Writing," and Appendix 2, "Technical Report Writing Checklist," for guidance on format and style for your report.

President Irene Bean wants a report summarizing this experiment. Your report, based on Sessions 4 and 12, should include:

1. Statement of problem and purpose of experiment.

2. Selection of response variable and the factors.

3. The importance of randomization and how it was accomplished.

4. A clear, complete, detailed description of your findings. You may include diagrams, tables, and graphs that help to clarify your description of the plant-growth experiment.

5. Recommendations, if any, about the best seed variety and watering scheme.

Name _____ Section _____ Session 12

SHORT ANSWER WRITING ASSIGNMENT

All answers should be complete sentences. Include all multiple scatter plots with this assignment.

1. Report the mean final heights of the plants for each treatment combination.

    Seed variety 1, water level 1 _____   Seed variety 1, water level 2 _____

    Seed variety 2, water level 1 _____   Seed variety 2, water level 2 _____

2. The first four multiple scatter plots contained three curves showing the growth of the three plants with the same treatment combination. Comment on the plant-to-plant variability for each graph.

3. Look at the multiple scatter plots of Amean and Cmean versus Week and Bmean and Dmean versus Week. Can you conclude that one seed variety grows better? Is this true for both of your water levels?

4. Look at the multiple scatter plots of Amean and Bmean versus Week and Cmean and Dmean versus Week. Can you conclude that one water level yields taller plants? Is this true for both of your seed varieties?

5. Can you conclude that one of the two factors in your study affects plant growth more than the other? Justify your answer.

6. Considering only the final heights, does it appear that one treatment combination produces the best plants? What are your recommendations to President Bean of Harvest Veg?

SESSION THIRTEEN

The Race to Solution

INTRODUCTION

Medical researchers use statistics to study the effect of a drug on the body. Often the study involves patients taking a new drug that previously has only been administered to lab animals. This raises important ethical questions. Which patients should receive the new drug? Although the new drug may be better than existing ones, side effects may be serious. Should one group of patients receive the treatment that is currently given, so comparisons can be made? How many patients should be used in the study? These questions illustrate only a few important issues that doctors, medical researchers, and biostatisticians must consider when they design an experiment to study the effects of medicine on the human body.

STATISTICAL CONCEPTS

Two-factor design, factor selection, treatments, interaction, descriptive statistics.

MATERIALS NEEDED

For each team, 10 tablets of each of two brands of noncoated aspirin, 20 clear plastic cups, a stopwatch, a 1/4-cup measure, a straightedge, a thin-tip marking pen, and 24 fluid ounces of both water and vinegar.

THE SETTING

You are a biostatistician working for Truth-B-Told, a consumer advocate organization that checks the validity of claims made by pharmaceutical companies. You have been asked to design a small study involving two brands of aspirin. The companies that make these aspirins both claim their product "brings quick relief to the common headache," but recent complaints from consumers have prompted your study.

BACKGROUND

For most types of medicine to have an effect, they must be absorbed into the bloodstream. Administering medicine by injection is the quickest way. However, this would be overkill when treating most minor aches and pains, so aspirin and other pain relievers are taken orally, in pill form. The medicine makes its way into your bloodstream as it dissolves in your stomach. With this in mind, it appears that a good aspirin, one that brings quick relief, is one that dissolves quickly. As human stomach juices are not available, we will study two different liquids that mimic what might be in our stomachs.

THE EXPERIMENT

STEP 1: FACTORS AND TREATMENT COMBINATIONS

In addition to the two brands of aspirin, your team will be provided two types of liquid. These are chosen to simulate certain stomach conditions. To clarify notation, we will call the liquids level 1 and level 2 and the brands of aspirin level 1 and level 2. Based on what your group will study, complete Table 13.1. We will refer to the combination of a level of the aspirin-brand factor and a level of the liquid factor as a **treatment combination**. There are four treatment combinations in this experiment.

STEP 2: RANDOMIZATION

While we attempt to control all factors in our experiment, there are some factors either that are beyond our control or that we are not aware of. To guard against one treatment combination getting an unfair advantage, we will use **randomization**. That is, we will use random numbers to decide the assignment of treatment combinations to cups. In this way all treatment combinations have an equal chance of being assigned to each cup, and no treatment combination has any advantage.

Table 13.1 Description of the Treatment Combinations

Notation	Generic Description	Detailed Description
(1, 1)	Level 1 of liquid and level 1 of aspirin	
(1, 2)	Level 1 of liquid and level 2 of aspirin	
(2, 1)	Level 2 of liquid and level 1 of aspirin	
(2, 2)	Level 2 of liquid and level 2 of aspirin	

We will use Minitab to generate the random numbers. These random numbers have the property that the numbers 1 and 2 are equally likely to occur at each point in the sequence. To generate these random numbers, launch Minitab and then:

1. Under the **Calc** menu, click and hold on **Random Data** and then select **Integer** from the submenu. An Integer Distribution dialog box similar to Figure 13.1 will appear.

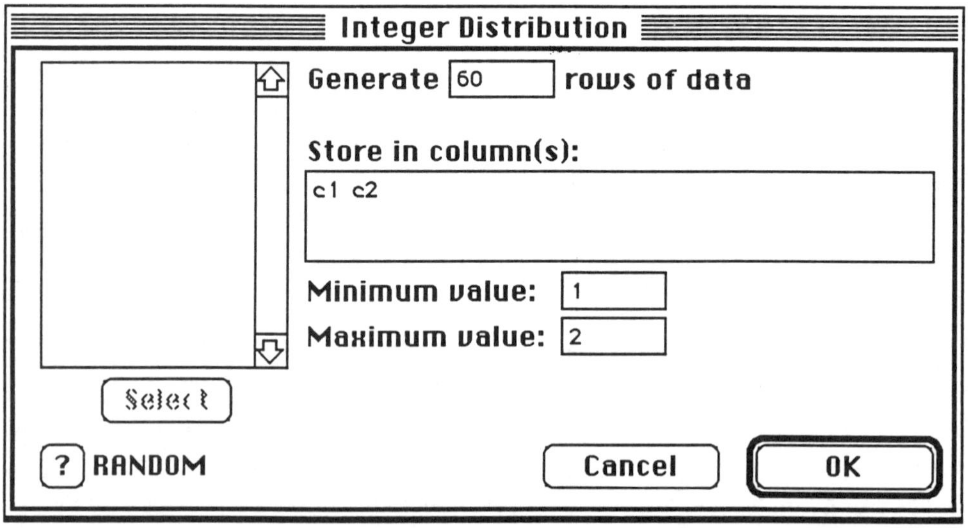

Figure 13.1 Integer Distribution Dialog Box

Table 13.2 Treatment Combination Assignments to Cups

Cup Number	Treatment Combination	Cup Number	Treatment Combination
	(1, 1)		(1, 2)
	(1, 1)		(1, 2)
	(1, 1)		(1, 2)
	(1, 1)		(1, 2)
	(2, 1)		(2, 2)
	(2, 1)		(2, 2)
	(2, 1)		(2, 2)
	(2, 1)		(2, 2)

2. Click in the box next to **Generate** and type the number of random numbers that you want. For this experiment 60 random numbers should be sufficient.

3. Click in the box beneath **Store in column(s)** and type **C1 C2**.

4. Click in the box to the right of **Minimum value** and type **1**.

5. Click in the box to the right of **Maximum value** and type **2**.

6. Click **OK**.

After a short delay, the random numbers will appear in columns C1 and C2 of the worksheet.

The first row of random numbers gives the treatment combination for the first cup. The level of the liquid factor is in C1 and the level of the aspirin-brand factor is in C2. After assigning a treatment combination to the first cup, go to the next row of random numbers and assign a treatment combination to the second cup, and so on. *Make sure you do not assign more than four cups to any treatment combination.* For example, if treatment combination (1, 1) already has four cups and the next row of random numbers is (1, 1), skip that row and go to the next row.

Each time you assign a treatment combination to a cup, write the cup number next to the treatment combination in Table 13.2. There is a small chance that you will run out of random numbers before you have assigned treatment combinations to all of the cups. If this happens, repeat the steps for generating random numbers. After Table

13.2 is completed, use the marking pen to neatly mark the *cup numbers* and *treatment combinations* on the 16 cups to be used in your study. The remaining 4 cups will be used for practice runs. Quit Minitab by selecting **Quit** under the **File** menu.

STEP 3: THE BASIC MEASUREMENT

Before your team can collect data, you need to make some basic decisions. How much liquid will be in the cup? The amount of liquid should be the same for all cups. We find that 2 fluid ounces (1/4 of a standard cup) works well with small plastic cups. A new cup and fresh liquid should be used with each aspirin.

The basic measurement is the time in seconds it takes for the aspirin to dissolve. The stopwatch is started the instant the aspirin enters the liquid. The stopwatch is stopped the instant the aspirin is "dissolved." Your team will have to decide what it means for an aspirin to be dissolved. You may decide that the aspirin is dissolved when it no longer retains a round shape. Alternatively, you may decide that it is dissolved when the largest remaining piece has a maximum dimension less than 1/8 of an inch. Waiting until no remaining aspirin pieces are visible will take a very long time.

The final decision your team needs to make is whether or not to shake the cups during the dissolving process. Shaking speeds the dissolving process. If you decide to shake, have a single person do the shaking and have him or her be as consistent as possible. For example, the person might gently shake the cup by moving it 2 inches, twice every 15 seconds.

To make these decisions and to perfect your technique, make one test run for each treatment combination using the unmarked cups. After the test runs are complete, your team should answer the following questions:

How much liquid will be poured into each cup?

What criterion is being used to decide when an aspirin is dissolved?

Table 13.3 Dissolving Times in Seconds

	Aspirin Brand 1		Aspirin Brand 2	
Liquid 1				
Liquid 2				

What method, if any, will be used to shake the cups?

STEP 4: DATA COLLECTION

We are now ready to collect the data. Begin with cup number 1. Use the levels of liquid and aspirin brand indicated on the cup. After measuring the dissolving time, record the result in Table 13.3. Repeat the process for cups 2–16, in order. Notice that we have four measurements, or **replicates**, for each treatment combination.

STEP 5: DATA ANALYSIS

We are now ready to enter the data. Launch Minitab. In the new worksheet, each row will correspond to a cup. Use column **C1** for the level (1 or 2) of the liquid factor for the cup, **C2** for the level of the aspirin-brand factor, and **C3** for the dissolving time. Name the columns using the variable names **Liquid** for **C1**, **Brand** for **C2**, and **Soltime** for **C3**. Soltime is short for the time required for the aspirin to dissolve into a solution. Figure 13.2 illustrates the Untitled worksheet at this point.

Enter the data from Table 13.3 into the worksheet. After entering the data for all cups, carefully double-check it. Then save the worksheet by selecting **Save Worksheet As** under the **File** menu. Name the worksheet **aspirin.data**.

The worksheet contains the treatment combination in two columns, C1 and C2. Let's create a new column that contains a two-digit number that describes each treatment combination. The tens digit of the two-digit number will indicate the level of

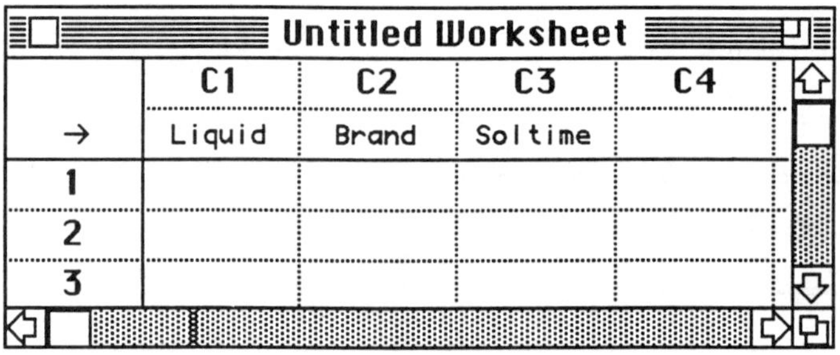

	C1	C2	C3	C4
→	Liquid	Brand	Soltime	
1				
2				
3				

Figure 13.2 Untitled Worksheet

the liquid factor and the units digit will indicate the level of the aspirin-brand factor. For example, this new variable will have the value 12 for treatment combination (1, 2), 21 for treatment combination (2, 1), and so on. This new variable will allow us to look at graphs and compute statistics for each of the four treatments separately. Give column **C4** the variable name **Combo**. Enter the two-digit treatment-combination number for each cup in column C4. After checking the values of the new variable, save the changes to your worksheet by selecting **Save Worksheet** under the **File** menu.

We begin our analysis by looking at dotplots of the data for each of the four treatment combinations. These dotplots will all have the same scale to make it easier to detect differences among the four groups.

To obtain dotplots using Minitab:

1. Under the **Graph** menu, select **Dotplot**. A Dotplot dialog box similar to Figure 13.3 will appear.

2. Click in the box under **Variables** and then double-click on **Soltime**.

3. Click in the box to the left of **By variable**. An **X** will appear in the box.

4. Click in the box to the right of **By variable** and then double-click on **Combo**.

5. Click **OK**.

The dotplots for the four treatment combinations will appear in the Session window. Get a printout of the dotplots by selecting **Print Window** under the **File** menu. Each value of Soltime is represented by a dot on one of the four graphs. Based on these simple graphs, what can you conclude about the effects, if any, of the two factors?

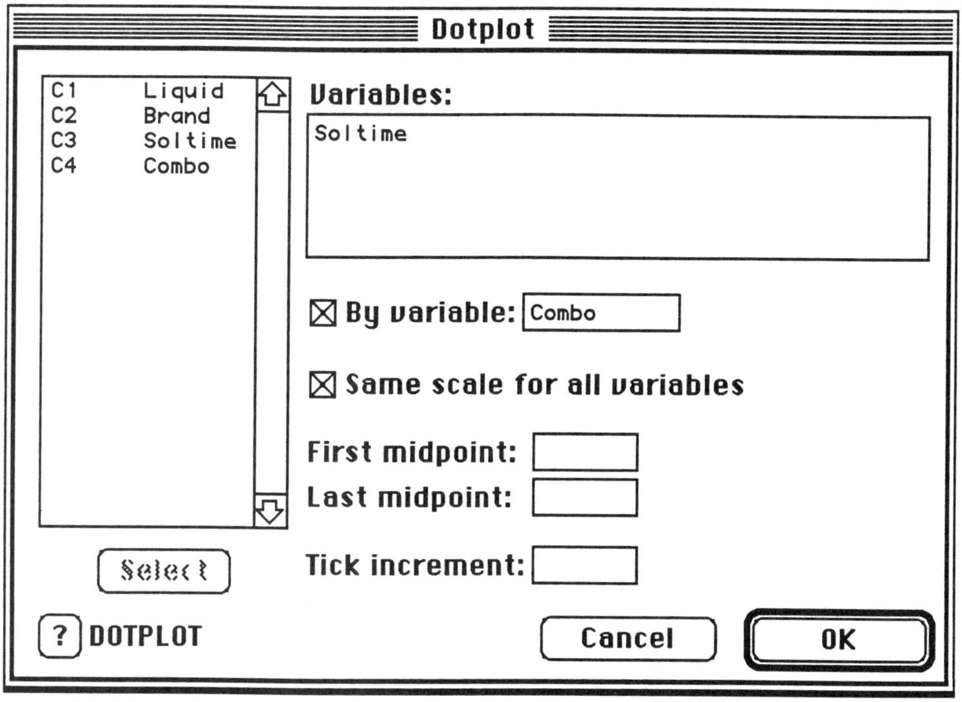

```
╔══════════════════ Dotplot ══════════════════╗
║  C1    Liquid   ⇧   Variables:              ║
║  C2    Brand        ┌─────────────────────┐ ║
║  C3    Soltime      │ Soltime             │ ║
║  C4    Combo        │                     │ ║
║                     │                     │ ║
║                     └─────────────────────┘ ║
║                                             ║
║                     ☒ By variable: [Combo]  ║
║                                             ║
║                     ☒ Same scale for all variables ║
║                                             ║
║                     First midpoint: [    ]  ║
║                     Last midpoint:  [    ]  ║
║                 ⇩                           ║
║      [ Select ]     Tick increment: [    ]  ║
║                                             ║
║  [?] DOTPLOT           [ Cancel ]  [  OK  ] ║
╚═════════════════════════════════════════════╝
```

Figure 13.3 Dotplot Dialog Box

Select **Data** under the **Window** menu to return to the worksheet. We will now compute descriptive statistics for each treatment combination.

1. Under the **Stat** menu, click and hold on **Basic Statistics** and then select **Descriptive Statistics** from the submenu. A Descriptive Statistics dialog box similar to Figure 13.4 will appear.

2. Click in the box under **Variables** and then double-click on **Soltime**.

3. Click in the box to the left of **By variable**. An **X** will appear in the box.

4. Click in the box to the right of **By variable** and then double-click on **Combo**.

5. Click **OK**.

Complete Table 13.4 by filling in the values of the mean ($\bar{X}$), median (M), and standard deviation (s) for each of the four treatment combinations.

The last part of our analysis is a graphical investigation of interaction. Suppose we consider only level 1 of the aspirin-brand factor and plot the mean dissolving time for each level of the liquid factor. An example graph is given in Figure 13.5. From this

Table 13.4 Summary Statistics for Dissolving Times in Seconds

	Aspirin Brand; Level 1	Aspirin Brand; Level 2
Liquid; Level 1	$\bar{X}=$ $M =$ $s =$	$\bar{X}=$ $M =$ $s =$
Liquid; Level 2	$\bar{X}=$ $M =$ $s =$	$\bar{X}=$ $M =$ $s =$

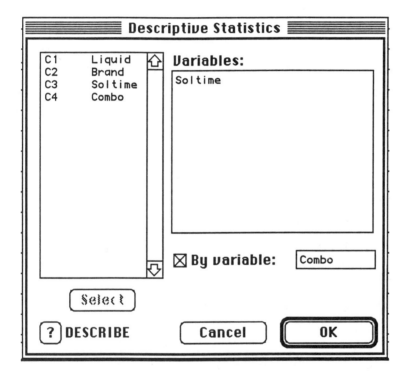

Figure 13.4 Descriptive Statistics Dialog Box

graph it is easy to see that for aspirin brand 1, the liquid factor influences the dissolving times. The aspirin takes twice as long to dissolve in liquid 2.

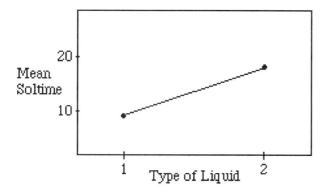

Figure 13.5 Plot of Mean Dissolving Time with Aspirin Brand 1 and Two Types of Liquid for Example Data

Now envision the same graph with the means also plotted for the other brand of aspirin. Such a graph would have two line segments, one for each brand. Two example graphs are given in Figure 13.6. In the left graph, the effect of the liquid factor on the dissolving times for both brands of aspirin is similar—liquid level 1 resulted in faster dissolving times for both types of aspirin. The parallel appearance of these two lines, indicating that the liquid factor has a similar effect on both types of aspirin brands, is an example of **no interaction**.

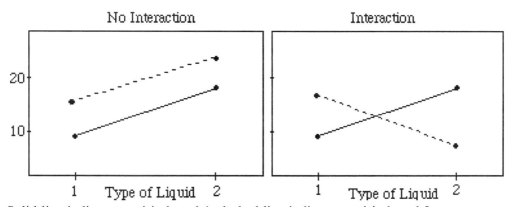

Solid line indicates aspirin brand 1; dashed line indicates aspirin brand 2

Figure 13.6 Plots Depicting No Interaction and Interaction of Factors

The right graph of Figure 13.6 shows a much different scenario. Note that there seems to be an opposite effect of the liquid factor for the two brands! This is an example of **interaction**. It is important to note that the two lines connecting the means do not have to cross to conclude interaction. We look for a highly nonparallel appearance of the lines in determining whether interaction exists.

The presence of interaction can lead to incorrect conclusions if the researcher is not careful. Consider the right graph of Figure 13.6 again. Suppose we averaged the two means for the two brands of aspirin and plotted them versus liquid level. Can you see that this plot would be a rather flat line (both means around 15), suggesting that the liquid factor does not affect dissolving times? However, this is not the case at all! The effect of the type of liquid on dissolving time very much depends on the brand of aspirin that is being used.

To produce a graph similar to Figure 13.6, we will create a new Minitab worksheet by selecting **New Worksheet** under the **File** menu. In this data set, we will type information on the sample mean dissolving times for the four treatment combinations. Use column **C1** to hold the liquid factor level, **C2** for the aspirin-brand factor level, and **C3** for the sample mean dissolving time. Name the columns using the variable name **Liquid** for **C1**, **Brand** for **C2**, and **Means** for **C3**. Enter the factor levels and the appropriate sample means from Table 13.4 in columns C1, C2, and C3. Double-check your entries. Figure 13.7 illustrates an untitled worksheet for some example data.

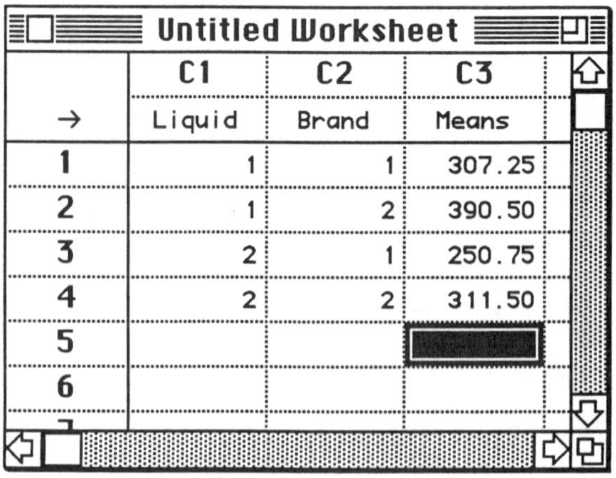

	C1	C2	C3
	Liquid	Brand	Means
1	1	1	307.25
2	1	2	390.50
3	2	1	250.75
4	2	2	311.50
5			
6			

Figure 13.7 Untitled Worksheet

To produce a plot investigating interaction:

1. Under the **Graph** menu, select **Scatter Plot**. A Scatter Plot dialog box similar to Figure 13.8 will appear.

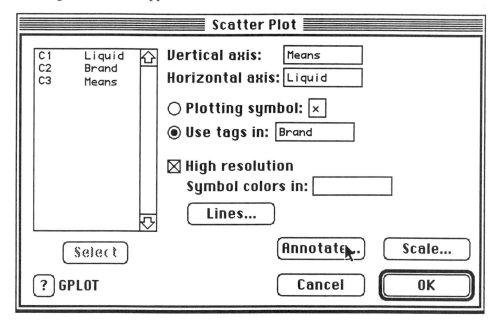

Figure 13.8 Scatter Plot Dialog Box

2. Click in the box to the right of **Vertical axis** and then double-click on **Means**.

3. Click in the box to the right of **Horizontal axis** and then double-click on **Liquid**.

4. To distinguish between the two brands, click on the circle next to **Use tags in**, click in the box to the right of **Use tags in**, and double-click on **Brand**.

5. Click **Annotate**. An Annotate Scatter Plot dialog box similar to Figure 13.9 will appear.

6. Type entries for the title, footnote, and axis labels in the appropriate boxes.

7. Click **OK** to close the Annotate Scatter Plot dialog box.

8. Click **OK** to end the Scatter Plot command.

The plotting symbol *A* marks the mean for brand 1 and the plotting symbol *B* marks the mean for brand 2.

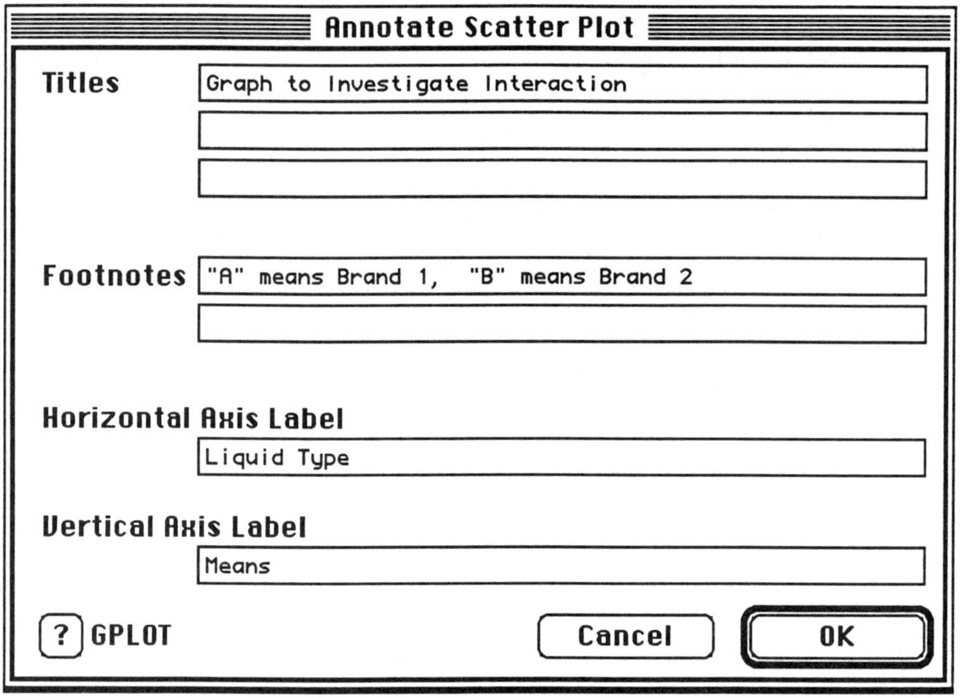

230 SESSION THIRTEEN

```
┌─────────────────── Annotate Scatter Plot ───────────────────┐
│                                                             │
│  Titles    ┌─────────────────────────────────────────────┐  │
│            │ Graph to Investigate Interaction            │  │
│            ├─────────────────────────────────────────────┤  │
│            │                                             │  │
│            ├─────────────────────────────────────────────┤  │
│            │                                             │  │
│            └─────────────────────────────────────────────┘  │
│                                                             │
│  Footnotes ┌─────────────────────────────────────────────┐  │
│            │ "A" means Brand 1,   "B" means Brand 2      │  │
│            ├─────────────────────────────────────────────┤  │
│            │                                             │  │
│            └─────────────────────────────────────────────┘  │
│                                                             │
│  Horizontal Axis Label                                      │
│            ┌─────────────────────────────────────────────┐  │
│            │ Liquid Type                                 │  │
│            └─────────────────────────────────────────────┘  │
│                                                             │
│  Vertical Axis Label                                        │
│            ┌─────────────────────────────────────────────┐  │
│            │ Means                                       │  │
│            └─────────────────────────────────────────────┘  │
│                                                             │
│  [?] GPLOT                         ( Cancel )   [  OK  ]    │
└─────────────────────────────────────────────────────────────┘
```

Figure 13.9 Annotate Scatter Plot Dialog Box

Print your graph by selecting **Print Window** under the **File** menu. You will need to use a straightedge to connect the means. Make sure you connect A to A and B to B, resulting in two lines on your graph. Take a look at your graph. Do you think interaction exists between the two factors?

This concludes our calculations on the data. Make sure you have all the output you need, and then quit Minitab by selecting **Quit** from the **File** menu.

PARTING GLANCES

We have only studied the effect of two factors on the dissolving times of aspirin. There are undoubtedly many other factors that affect the dissolving time of medicine in our stomachs. Temperature of the liquid may be a significant factor. The amount of undigested food in the stomach is another. Can you see how these would have been more difficult to control in our experiment?

We have used descriptive statistics and plotting techniques to analyze the dissolving times. While these methods are informative, they are not entirely satisfactory. They do not tell us anything about the probability of observing the differences in dissolving times, if in fact the brand or liquid factors had no effect. To find these probabilities, we would use a form of inferential statistics known as **analysis of variance**, which is beyond the scope of this discussion.

In our experiment we have considered noncoated aspirins. It is likely that a pill with a coating will take longer to dissolve; however, it is well known that such a coating helps to prevent the formation of ulcers in the stomach. For this type of aspirin, it would be interesting to study and quantify the trade-off between a longer absorption time and a lower risk of ulcers. Clearly this would require a much more elaborate experimental design with actual clinical trials involving patients taking the drugs.

EXTENDED WRITING ASSIGNMENT

Refer to Appendix 1, "Technical Report Writing," and Appendix 2, "Technical Report Writing Checklist," for guidance on format and style for your report.

Dr. Ruth Nadir, president of Truth-B-Told, would like a report summarizing your findings. The report should include:

1. A summary of the experiment, including discussion of why this experiment was performed

2. A brief description of the factor levels, the measurement of interest, and the randomization that was used

3. A discussion of the effects of the factors on the dissolving times, including interaction if present

4. Recommendations to Dr. Nadir concerning the validity of the manufacturers' claims

5. Limitations of the current experiment and ideas for future experimentation

Name _____ Section _____ Session 13

SHORT ANSWER WRITING ASSIGNMENT

All answers should be complete sentences. Include a copy of Table 13.4 and your interaction graph with this assignment.

1. What were the two levels of each factor?

2. Was randomization used in this experiment? Why?

3. Based on the values in Table 13.4, do you think the level of the liquid factor affected dissolving times? If so, which liquid resulted in faster dissolving times?

4. Based on the values in Table 13.4, do you think the brand of aspirin affected dissolving time? If so, what brand do you recommend?

5. Based on your graph investigating interaction, do you think the two factors interact? Explain.

6. Of all four treatment combinations, did one combination clearly yield the fastest dissolving time?

SESSION FOURTEEN

Walk This Way

INTRODUCTION

Power-walking is an important form of aerobic exercise, especially for middle-aged and elderly individuals. What are the effects of different walking styles on walking speed and heart rate? Research questions like these are typical to the field of exercise physiology, where formal experimental design and statistical analysis have been instrumental in important discoveries made in recent years.

STATISTICAL CONCEPTS

Blocking, paired-sample t confidence intervals and hypothesis tests.

MATERIALS NEEDED

For the class: three stopwatches; a 12-inch bowl; a coin; and slips of paper marked A, B, C, ... , O.

THE SETTING

We will compare two walking styles:

1. Walking with arms at sides
2. Exaggerated arm movement walking

We will compare these styles in terms of time required to walk a fixed distance (walk time) and heart-rate increase. For the arms-at-sides walking style (see Figure 14.1), walk with both arms held firmly at your side, as if your thumbs were hooked in your front pants pockets. For the exaggerated-arm-movement style (see Figure 14.2), walk like a toy soldier, raising each arm to a horizontal position in front of you about shoulder high as you step with the opposite leg; at the same time, raise your other arm behind you as a counterbalance. Your instructor will demonstrate, but don't laugh too much—your turn is coming!

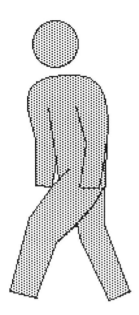

Figure 14.1 Arms-at-Sides Walking Style

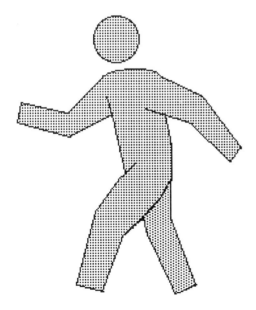

Figure 14.2 Exaggerated-Arm-Movement Walking Style

BACKGROUND

There will be a great deal of variability among individuals for both the heart-rate increase and the walk-time measurements. Considering walk time as an example, some people just naturally walk faster than others. One way to compare the two walking styles would be to divide our participants at random into two groups and let each group test a different walking style. That study design would be an **independent samples** design, since no one person belongs to both test groups. If we were to do the study this way, by chance one group might get more than its share of fast walkers. The comparison would still be valid (with the right analysis), but it would not be efficient. Because so much variability exists among individuals, we would need very large groups to get an accurate comparison of the population mean walk times for these two walking styles.

Instead of dividing the class into two groups, each participant will test both walking styles, in a random order. We will thus obtain two walk times on each subject, and the subject-to-subject variability will be removed from the analysis by taking the difference between these two walk times. Then a one-sample statistical analysis will be done on these differences. The technique of removing subject-to-subject variability, one of the most important concepts in experimental design, is known as **blocking** or **local control**.

Participation in the walking exercise of this experiment is completely voluntary. If you are uncomfortable in any way with the thought of participating, by all means feel free to decline. Minor risks are involved, such as tripping, so your teacher may provide a waiver form for your signature. Be careful!

THE EXPERIMENT

STEP 1: THE BASIC MEASUREMENTS

Each participant will, in turn:

1. Count his or her 15-second heart rate.

2. Walk the distance, about 50 yards, *as fast as is safely possible* using either the arms-at-sides style or the exaggerated-arm-movement style (as determined by a coin flip).

3. Count the 15-second heart rate again.

Each walk will thus produce three measurements:

1. The heart rate before walking (abbreviated Hrb)

2. The walk time

3. The heart rate after walking (Hra)

After all participants have walked once, all will walk again in the same order, with each person using the walking style he or she did not use the first time.

Remember, walk as quickly as you safely can for the entire distance. Don't slow down at the end and don't favor one walking style over the other. If you do, you might bias the study.

STEP 2: CONDUCT THE EXPERIMENT

The class will proceed to a predetermined walking area. Three class members will be needed to keep time with the stopwatches, and a fourth (the statistician) to record the results. Ten to fifteen volunteer walkers are ideal.

Each participant will be randomly assigned a letter—A, B, C, and so on— by choosing slips of paper from a bowl. These letters determine the order of walking. Before each participant's first walk, a coin will be flipped to determine which walking method he or she will use first. To be specific, if your initial coin flip is:

Heads: your first walk is arms-at-sides

Tails: your first walk is exaggerated-arm-movement

The statistician will record the heart rates and walk times in his or her Table 14.1.

Table 14.1 Walking Experiment Raw Data

Subject Name	Subject Letter	Arms-at-Sides			Exaggerated-Arm-Movement		
		Heart Rate Before	Walk Time	Heart Rate After	Heart Rate Before	Walk Time	Heart Rate After
	A						
	B						
	C						
	D						
	E						
	F						
	G						
	H						
	I						
	J						
	K						
	L						
	M						
	N						
	O						

STEP 3: DATA ENTRY AND MANIPULATION

When all participants have walked twice, return to the laboratory. The instructor will read aloud the experimental results. Notice that walkers with small walk times under one walking style also tend to have small walk times under the other style. Choose a Macintosh and launch Minitab. In the new worksheet, name the first variable **Subject** (these will be the subject letters). Name the other columns as follows:

Hrb1   (heart rate before arms-at-sides walking)

Wlktime1   (walk time for arms-at-sides walking)

Hra1   (heart rate after arms-at-sides walking)

Hrb2   (heart rate before exaggerated-arm-movement walking)

Wlktime2   (walk time for exaggerated-arm-movement walking)

Hra2   (heart rate after exaggerated-arm-movement walking)

Such cryptic variable names may be annoying, but Minitab will not allow more than eight characters in a variable name. Figure 14.3 illustrates the variable names in the Untitled worksheet.

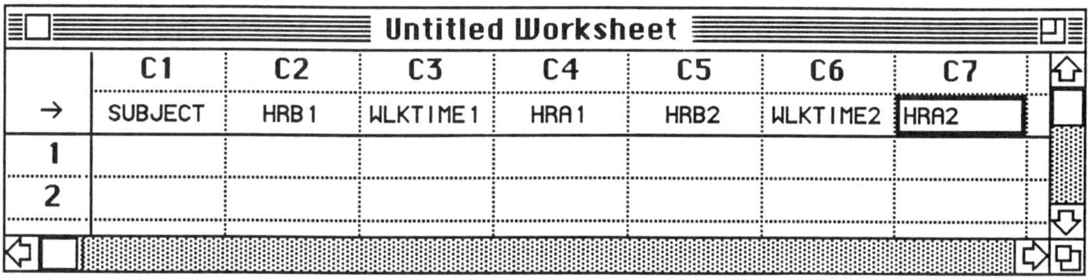

	C1	C2	C3	C4	C5	C6	C7
→	SUBJECT	HRB1	WLKTIME1	HRA1	HRB2	WLKTIME2	HRA2
1							
2							

Figure 14.3 Untitled Worksheet

Begin entering the raw data from Table 14.1 into the worksheet. When you have finished and have double-checked for errors, save the data set onto your diskette by selecting **Save Worksheet As** under the **File** menu. Use the name **Walkdata**.

Now, create the variable **Rateinc1** to reflect increase in heart rate per minute ("after" rate - "before" rate) for walking style 1, arms at sides, as follows:

1. Under the **Calc** menu, click and hold on **Functions and Statistics** and then select **General Expressions** from the submenu. A General Expressions dialog box similar to Figure 14.4 will appear.

2. Click in the box next to **New/modified variable** and type **Rateinc1**.

3. Click in the box under **Expression** and type **4*('Hra1'-'Hrb1')**. This computes the increase in the 15-second heart rate and converts to beats per minute.

4. Click **OK**.

After a moment the data for the new variable should appear in the worksheet.

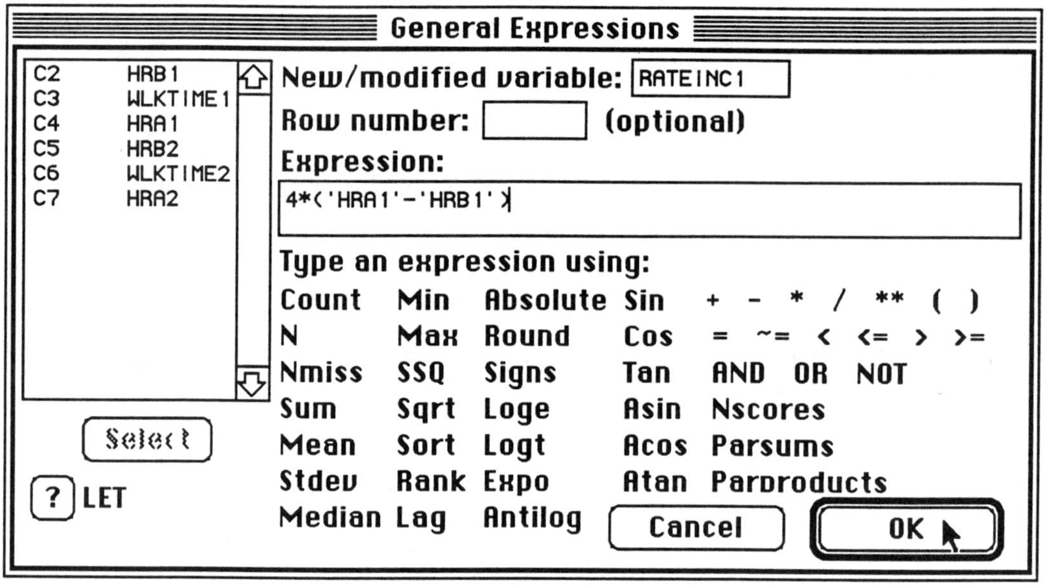

General Expressions

C2 HRB1	New/modified variable: `RATEINC1`
C3 WLKTIME1	Row number: ☐ (optional)
C4 HRA1	Expression:
C5 HRB2	`4*('HRA1'-'HRB1')`
C6 WLKTIME2	
C7 HRA2	

Type an expression using:

Count	Min	Absolute	Sin	+ − * / ** ( )
N	Max	Round	Cos	= ~= < <= > >=
Nmiss	SSQ	Signs	Tan	AND OR NOT
Sum	Sqrt	Loge	Asin	Nscores
Mean	Sort	Logt	Acos	Parsums
Stdev	Rank	Expo	Atan	Parproducts
Median	Lag	Antilog		

[Select] [?] LET   [Cancel] [OK]

Figure 14.4 General Expressions Dialog Box

Now carry out similar steps, making the necessary minor changes to commands, to create **Rateinc2**, the measure of heart-rate increase for each subject under walking style 2, exaggerated-arm-movement.

Next, we will create a variable to compare the walk time of the arms-at-sides style to that of the exaggerated-arm-movement style, within each participant. We'll call it **Dwlktime** (difference in walk times). It is defined by

Dwlktime = (arms-at-sides walk time) −
(exaggerated-arm-movement walk time).

1. Under the **Calc** menu, click and hold on **Functions and Statistics** and then select **General Expressions** from the submenu. A General Expressions dialog box will appear.

2. Click in the box next to **New/modified variable** and type **Dwlktime**.

3. Click in the box under **Expression** and type **'Wlktime1'-'Wlktime2'**.

4. Click **OK**.

If a participant has a positive value for Dwlktime, the exaggerated-arm-movement walking style was faster for him or her, because the arms-at-sides walking style took more time than did the exaggerated-arm-movement walking style. Keep this fact in mind throughout the rest of analysis: Positive values of Dwlktime suggest that the exaggerated-arm-movement style is a faster walking style.

The last variable we will create is a variable to compare the heart-rate increase under the arms-at-sides walking style to the increase under the exaggerated-arm-movement walking style, within each participant. We'll call it **Drateinc** (difference in heart-rate increase), and it is defined by

Drateinc = (arms-at-sides heart-rate increase) -
    (exaggerated-arm-movement heart-rate increase).

1. Under the **Calc** menu, click and hold on **Functions and Statistics** and then select **General Expressions** from the submenu. A General Expressions dialog box will appear.

2. Click in the box next to **New/modified variable** and type **Drateinc**.

3. Click in the box under **Expression** and type **'Rateinc1'-'Rateinc2'**.

4. Click **OK**.

Save the additions you have made to the data set at this time by selecting **Save Worksheet** under the **File** menu.

If a participant has a positive value for Drateinc, the arms-at-sides walking style produced a greater increase in his or her heart rate than the exaggerated-arm-movement walking style. Keep this fact in mind throughout the rest of the analysis: A positive value of Drateinc suggests that the arms-at-sides style had a stronger effect on the participant's heart than the exaggerated-arm-movement style.

STEP 4: DESCRIPTIVE STATISTICS AND PLOTS

Obtain standard descriptive statistics for all the numeric variables:

1. Under the **Stat** menu, click and hold on **Basic Statistics** and then select **Descriptive Statistics** from the submenu. A Descriptive Statistics dialog box similar to Figure 14.5 will appear.

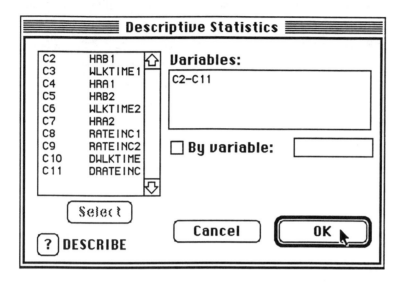

Figure 14.5 Descriptive Statistics Dialog Box

2. Click in the box under **Variables** and type **C2-C11**.

3. Click **OK**.

After a moment descriptive statistics for all the variables will appear in the Session window.

Next we will obtain graphical summaries for the comparison variable Dwlktime and its component parts Wlktime1 and Wlktime2. We will obtain side-by-side dotplots of these three variables to facilitate comparisons.

1. Under the **Graph** menu, select **Dotplot**. A Dotplot dialog box similar to Figure 14.6 will appear.

2. Click in the box under **Variables** and then double-click on the variable names **Wlktime1**, **Wlktime2**, and **Dwlktime**.

3. Click in the box next to **Same scale for all variables**.

4. Click **OK**.

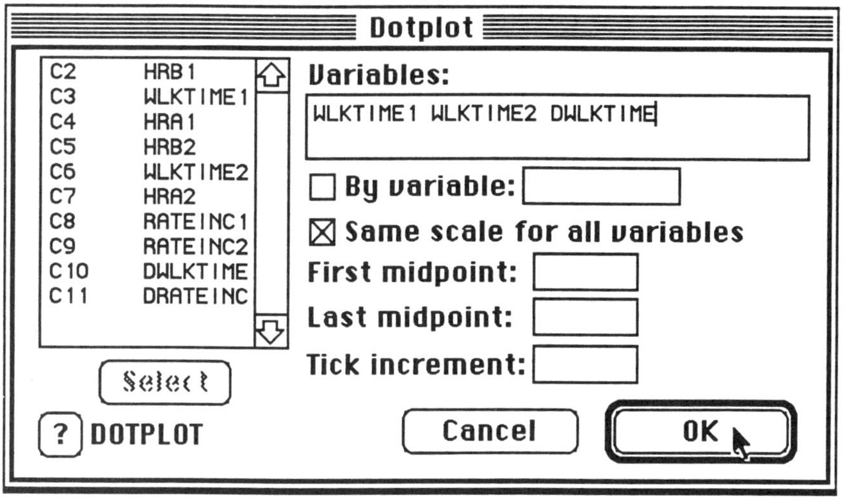

Figure 14.6 Dotplot Dialog Box

Now repeat the above construction of dotplots using the three heart-rate increase variables Rateinc1, Rateinc2, and Drateinc.

STEP 5: STATISTICAL INFERENCE

It would be nice if the entire population of all students could be tested to compare the two walking styles. We only tested a sample group of students, but (if it was randomly selected) formal statistical techniques based on probability allow us to make careful **confidence statements** about what the difference in population means would have been if we could have tested the entire population. Of course, our test group was not really randomly selected from the population of all students. We will assume it was, for the sake of example (this would not be acceptable for serious research).

To formally compare the arms-at-sides walking style (treatment 1) to the exaggerated-arm-movement style (treatment 2), we will compute confidence intervals for the population means of Dwlktime and Drateinc, and later test hypotheses on these means, using Student's t distribution. It is important that the distributions of Dwlktime and Drateinc are approximately normal in order for the t-tests and confidence intervals to be valid. To obtain 90 percent confidence intervals:

1. Under the **Stat** menu, click and hold on **Basic Statistics** and then select **1-Sample t** from the submenu. A 1-Sample *t* dialog box similar to Figure 14.7 will appear.

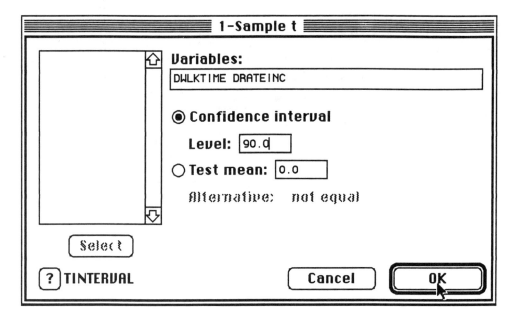

Figure 14.7 1-Sample t Dialog Box

2. Click under **Variables** and then double-click on **Dwlktime** and **Drateinc**.
3. Click in the box next to **Level** and type **90**.
4. Click **OK**.

The confidence intervals should appear in your Session window and at the printer. What do these intervals mean? For example, for Dwlktime we can state with 90 percent confidence that, if the entire population of students could have been tested, the mean of their walk-time differences would lie in the computed interval. For our purposes, "90 percent confidence" means we would be willing to bet up to $9 against $1 that this statement is true. As

(population mean Dwlktime) = (population mean arms-at-sides walk time) -
(population mean exaggerated-arm-movement walk time),

Table 14.2 Example Confidence Intervals for Mean Dwlktime with Suggested Interpretations

90% Interval	Suggested Interpretation
(2, 6)	The population mean arms-at-sides walk time is at least 2 and at most 6 seconds higher than the population mean exaggerated-arm-movement walk time.
(-20, -12)	The population mean exaggerated-arm-movement walk time is at least 12 and at most 20 seconds higher than the population mean arms-at-sides walk time.
(-2, 1)	It is not clear which walking style has a larger population mean walk time. If the arms-at-sides style has a larger population mean walk time, it would be by at most 1 second. If the exaggerated-arm-movement method has the larger population mean walk time, it would be by at most 2 seconds.
(-10, 20)	It is not clear which walking style has a larger population mean walk time. If the arms-at-sides style has a larger population mean walk time, it would be by at most 20 seconds. If the exaggerated-arm-movement method has the larger population mean walk time, it would be by at most 10 seconds.

we can use the confidence interval to make statements about which walking style would have a smaller population mean walk time. Table 14.2 shows several possible confidence intervals that we might obtain and suggested interpretations. Of course, similar interpretations would be used for heart-rate increase confidence intervals.

Be very careful with your interpretations. Your experiment may have been a success, but if you misinterpret the results you may do more harm than good. Note in particular that just because the confidence interval contains zero doesn't mean we have nothing to say. The third and fourth intervals contain zero, but in the third case we can say that the population mean walk times would be very close to each other, whereas in the fourth case (if we require 90 percent confidence in our statement) we can't say anything useful. Wide confidence intervals occur when the number of participants in the sample is too small to get an accurate comparison of the walking styles.

Tests of hypotheses are another form of formal statistical inference. For walk time, we would like to test the null hypothesis that there is no difference between population mean walk times for the two walking styles:

H_0: population mean arms-at-sides walk time =
population mean exaggerated-arm-movement walk time

Should the alternative hypothesis be "greater than," "less than," or "not equal"? This should be decided without looking at the data; ideally, it should be decided before the data is even collected. In our setting, it is not clear before the experiment what effect the exaggerated arm movement should have on walk time, so we should probably use a two-sided, "not equal," alternative hypothesis:

H_A: population mean arms-at-sides walk time ≠
population mean exaggerated-arm-movement walk time

Now, since

(population mean Dwlktime) = (population mean arms-at-sides walk time) -
(population mean exaggerated-arm-movement walk time),

the above hypotheses are equivalent to

H_0: population mean Dwlktime = 0

H_A: population mean Dwlktime ≠ 0

We can test these using our sample of Dwlktimes:

1. Under the **Stat** menu, click and hold on **Basic Statistics** and then select **1-Sample t** from the submenu. A 1-Sample *t* dialog box similar to Figure 14.8 will appear.

2. Click in the box under **Variables** and then double-click on **Dwlktime**.

3. Click in the circle next to **Test mean**.

4. Since the default alternative hypothesis is **not equal**, we do not need to change this for **Dwlktime**.

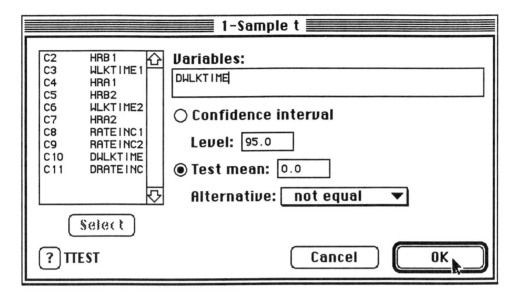

Figure 14.8 1-Sample t Dialog Box

5. Click **OK**.

The t-test results will appear in the Session window and at the printer. Of special importance is the *p*-value, also called the **observed significance level** of the test. In hypothesis testing, we decide to reject the null hypothesis H_0 if the data we actually observed would have been highly unlikely to occur if H_0 were true. The *p*-value tells us exactly how unlikely (in support of the alternative hypothesis) the observed data would be if the null hypothesis were true. For example, if the *p*-value = .03, in only 3 percent of experiments like this would we observe sample means as different as or more different than the ones we got, if in fact H_0 were true. Since 3 percent is pretty rare, *p*-value = .03 would lead most people to doubt the truth of the null hypothesis. The smaller the *p*-value is, the stronger is the evidence against H_0 in the data. Everyone gets to select the cutoffs, but most researchers have opinions similar to these:

If the *p*-value is less than .05, there is strong evidence against H_0, so assert that H_A holds.

If the *p*-value is between .05 and .15, there is some evidence (not overwhelming) against H_0.

If the *p*-value is larger than .15, there is little or no evidence against H_0.

Notice that we never say there is strong evidence for H_0. This is akin to a courtroom trial, where the jury might find the defendant not guilty but would not pronounce him or her innocent.

Use the *p*-value for your hypothesis test to decide which of the above categories applies. Then, in writing the interpretations, substitute meaningful words for the symbols H_0 and H_A. For example, if the *p*-value = .003, this is very strong evidence against H_0, so assert H_A. That is, we assert that the population mean walk times would be different for the two walking styles.

For testing hypotheses on mean heart-rate increase between the two walking styles, the null hypothesis should again be that the population mean Drateinc is 0. It seems in this case that the correct alternative hypothesis is "less than," since we expect the exaggerated-arm-movement style to put more strain on the heart. That is, we expect

> population mean arms-at-sides heart-rate increase <
> population mean exaggerated-arm-movement heart-rate increase

Therefore we expect population mean Drateinc = Rateinc1 - Rateinc2 < 0. Formally, we should test

H_0: population mean Drateinc = 0

H_A: population mean Drateinc < 0

To test these, repeat the hypothesis test steps above, except double-click on **Drateinc** in item 2, and in item 4 click and hold the arrow next to **not equal**. In the menu box that appears, select the **less than** alternative hypothesis.

This concludes our calculations on the data. If the Session window is not currently active, select **Session** under the **Window** menu. Print your descriptive statistics, dotplots, confidence intervals, and hypothesis test results by selecting **Print Window** under the **File** menu. After you get your output, quit Minitab by selecting **Quit** under the **File** menu.

PARTING GLANCES

We have compared the two walking styles using a paired or dependent sample experiment: Every subject walked both styles. Would it have been better to have used an independent samples approach as described earlier, in the "Background" section? We said that this is really a question of efficiency: Which strategy requires fewer subjects to obtain an accurate comparison of the walking styles? To be more precise, which strategy requires fewer subjects to obtain a confidence interval of a specified width for the difference of population mean walk times? We can use our paired-sample data to answer this question. Define

s_D^2 = sample variance (= standard deviation squared) of the Dwlktime values

s_1^2 = sample variance of the Wlktime1 values

s_2^2 = sample variance of the Wlktime2 values

You have all these values in your descriptive statistics output. We will define the **relative efficiency** of the paired-sample approach to the independent-samples approach by

$$R = \frac{2(s_1^2 + s_2^2)}{s_D^2}$$

R is approximately the ratio

$$\frac{\text{total number of subjects (independent samples)}}{\text{total number of subjects (paired sample)}}$$

needed to obtain confidence intervals for the difference in population means of equal width. So, if $R = 1.5$, we would need 50 percent more subjects using the independent-samples strategy to get a confidence interval of the same width as the paired-sample strategy. If $R = 6$, we would need 6 times as many subjects! Of course, under the paired-sample approach, each subject does twice as much work. In the assignment for this session, you are asked to compute and interpret the relative efficiency ratio R for the data we collected in this experiment.

EXTENDED WRITING ASSIGNMENT

Refer to Appendix 1, "Technical Report Writing," and Appendix 2, "Technical Report Writing Checklist," for guidance on format and style for your report.

Write a report summarizing the "Walk This Way" experiment and results. It should include a complete description of the experiment, summary of the data analysis, and presentation of the data. In the discussion, you should also include:

1. Careful interpretations of the confidence intervals computed on walk times and heart-rate increases

2. Careful interpretations of the hypothesis tests carried out on walk times and heart-rate increases

3. Computation and interpretation of the relative efficiency ratio R

Name _____ Section _____ Session 14

SHORT ANSWER WRITING ASSIGNMENT

All answers should be complete sentences. Include your six dotplots with this assignment.

1. Write the descriptive statistics in Table 14.3.

Table 14.3 Descriptive Statistics for Walking Experiment ($n = $ ___)

Variable	Mean	Std. Dev.	Variance
Wlktime1			
Wlktime2			
Dwlktime			
Rateinc1			
Rateinc2			
Drateinc			

2. From your dotplots, how would you describe the shape of the distribution of each variable listed in Table 14.3 (e.g., skewed, symmetric, bimodal, outliers)?

3. Give and carefully interpret your confidence interval for population mean Dwlktime. Use the examples in Table 14.2 as a guide.

4. Give and carefully interpret your confidence interval for population mean Drateinc.

5. Give and carefully interpret the *p*-value for the hypothesis test you carried out on population mean Dwlktime.

6. Give and carefully interpret the *p*-value for the hypothesis test you carried out on population mean Drateinc.

7. Compute and interpret the relative efficiency ratio R defined in the "Parting Glances" section for the walk-time variable.

SESSION FIFTEEN

Absorbency of Paper Towels—A Messy Data Problem

INTRODUCTION

We often hear about the quick absorbency of a particular brand of paper towel. But did you ever wonder whether the store-brand of paper towel is as good as the big-name, big-advertising-budget, big-price brand?

STATISTICAL CONCEPTS

Population, census, simple random sample, systematic random sample, side-by-side boxplots, independent samples t-test and confidence interval.

MATERIALS NEEDED

For each team, a balance sensitive to the nearest gram, one roll of name-brand paper towels, one roll of store-brand paper towels having the same dimensions as the name-brand towels, a 5-quart bucket of water, a 12-inch bowl, a coin, tongs, a watch with a second hand or a stopwatch, and a thin-tip marking pen.

WEIGHING WITH A DIGITAL ELECTRONIC BALANCE

Most balances can weigh in either grams or ounces. Set your balance to weigh in grams if possible. Make sure the balance is on a level surface and turn it on. When zero is displayed, the balance is ready for use. Place the object to be weighed on the platform of the balance and read the displayed weight.

THE SETTING

You are commissioned by a consumer magazine to investigate a claim made by a grocery store that its store-brand paper towels are as absorbent as the name-brand towels. It's a messy job, but someone has to do it. On the bright side, the subjects in this experiment do double duty as cleanup materials.

BACKGROUND

Obtaining a "random sample" from the output of an industrial process is more easily said than done. In this example, we would like random samples of towels from the population of those made by the name-brand company and from the store brand. The best we can do here is to obtain a roll of each brand and then test a random sample of several sheets from each roll. Of course, the tested sheets on a particular roll do not, strictly speaking, comprise a random sample from the population of all sheets of that brand.

THE EXPERIMENT

STEP 1: THE BASIC MEASUREMENT

We will test the absorbency of a paper towel as described below. One team member needs to keep track of time, while the other does the following:

1. Select a paper towel and fold it twice, to one-fourth its original size.
2. Put the dry paper towel in the dry bowl, put the bowl on the balance, and record the total weight. The combined weight of the bowl and the dry sheet will be called the **dry weight**.
3. Holding the sheet in the tongs, immerse the sheet completely in the bucket of water for 10 seconds.
4. With the tongs, hold the wet sheet above the water, letting it drip for 20 seconds. While doing this, remove the bowl from the balance and reset the balance.
5. Place the wet sheet back in the bowl, place the bowl and sheet again on the balance, and record the total weight. The combined weight of the bowl and the wet towel will be called the **wet weight**.
6. Discard the wet sheet and dry the bowl with an extra paper towel.

The weight of the water absorbed by the paper towel is the wet weight minus the dry weight. Try obtaining this measurement on a sheet or two for practice.

STEP 2: SELECTING THE SAMPLES

As you know by now, much of the business of statistics is involved in attempting to gain reliable information about a population of subjects or experimental units. In this setting, each roll of sheets is a population. An "overkill" experiment might involve testing every sheet on each roll. If we examine every unit in the population, we are performing a **census**. If we census these populations, we could use the mean weight of water absorbed by all towels on the roll, the **population mean**, as a measure of performance. Let us assume that the central goals of our experiment are to accurately estimate the population mean water absorbed per sheet for each of the two rolls and to compare the estimates.

We will not census these populations. There is not enough time to test every sheet on both rolls. It is also unnecessary, since carefully taken samples of sufficient size will provide very accurate estimates of population means. We will test a sample of sheets from each roll and use the sample mean weight of water absorbed per sheet as an estimate of the population mean.

Suppose for the sake of this discussion that we decide to choose a sample of 10 sheets from each roll. How should we select these test sheets? Here are three possibilities for a roll of 100 sheets.

1. Convenience sample: We could take the first 10 sheets on each roll.

2. Systematic random sample: We could randomly choose 1 of the first 10 sheets to test and then carefully disassemble the entire roll, taking every 10th sheet after the first sheet.

3. Simple random sample: We could carefully disassemble the entire roll, marking a number on each sheet. Then, we could randomly choose 10 numbers between 1 and 100, and these would determine which sheets would be tested.

The instructor will lead a discussion on the questions, "What can cause a convenience sample to be unrepresentative of the roll? How about a systematic sample or a simple random sample?"

No sampling method can guarantee a sample mean that accurately estimates the population mean. If we use a convenience sample, we can only cross our fingers and hope that the first sheets on the roll are not different in some way from the rest of the roll. Systematic random sampling is widely used and gives each sheet on the roll an equal chance of being selected for the sample. However, there is no way under systematic random sampling to say with any confidence how accurate the sample mean is as an estimate of the population mean. If we use a simple random sample, we still cannot guarantee an accurate estimate. However, we can say with high confidence how accurate the sample mean is by forming a confidence interval for the population mean. In a sense, the simple random sample harnesses chance and gets it to work for us.

So, we will select simple random samples of sheets to test. Ten sheets will be randomly selected from each roll. First, prepare the population:

1. Carefully tear off all the remaining sheets on the roll. Set aside any sheets that are severely damaged in the process. Don't get in a hurry—do it right.

2. With a soft-tip felt marking pen, number the undamaged sheets with a small number in the upper right corner of each sheet.

When this is completed, use Minitab to generate 20 random integers between one and N, where N is the number of undamaged sheets in your name-brand population. To do this, launch Minitab and:

1. Under the **Calc** menu, click and hold on **Random Data** and then select **Integer** from the submenu. An Integer Distribution dialog box similar to Figure 15.1 will appear.

2. Click in the box to the right of **Generate** and type **20**.

3. Click in the box under **Store in column(s)** and type **C1**.

4. Click in the box to the right of **Minimum value** and type **1**.

5. Click in the box to the right of **Maximum value** and type your value of N.

6. Click **OK**.

After a moment, numbers between 1 and N should appear in column C1. Write the first 10 *nonrepeated* randomly generated numbers in the space below, and select these sheets from the name-brand population for absorbency testing.

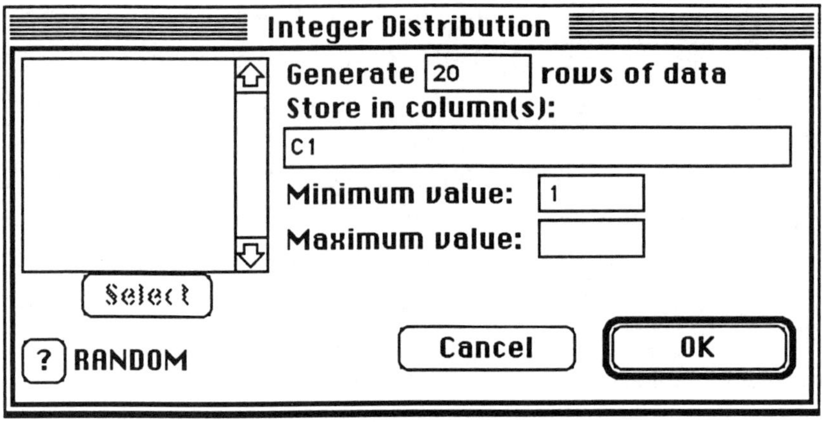

Figure 15.1 Integer Distribution Dialog Box

Repeat the above steps to prepare the store-brand roll and randomly choose test sheets from it. You must obtain a different set of 20 random numbers for the store-brand roll. Remember that N is now the number of undamaged store-brand sheets in your population. Write the first 10 nonrepeated randomly generated numbers in the space below, and select these sheets from the store-brand population for testing.

Quit Minitab by selecting **Quit** under the **File** menu.

STEP 3: CONDUCTING THE EXPERIMENT

After ten sheets from each roll have been selected, work in teams to carefully make the absorbency measurements described in Step 1. Flip a coin before each test to randomly choose which brand to test. If the coin lands heads, test the next name-brand towel and record its measured dry and wet weights in the first available row labeled Brand 1 (rows 1–10) of Table 15.1. If the coin lands tails, test the next store-brand towel and record its measured dry and wet weights in the first available row labeled Brand 2 (rows 11–20) of Table 15.1. Of course, when all ten towels of one brand have

Table 15.1 Paper-Towel Absorbency Raw Data in Grams for Name-Brand (Brand 1) and Store-Brand (Brand 2) Towels

Test	Brand	Dry Weight	Wet Weight
1	1		
2	1		
3	1		
4	1		
5	1		
6	1		
7	1		
8	1		
9	1		
10	1		
11	2		
12	2		
13	2		
14	2		
15	2		
16	2		
17	2		
18	2		
19	2		
20	2		

been tested, you don't need to flip the coin any more. In the Dry Weight column of Table 15.1, record the measured weight of the dry sheet and bowl together. In the Wet Weight column, record the measured weight of the wet sheet and bowl together.

STEP 4: DATA ANALYSIS

Choose a Macintosh and launch Minitab. In the new worksheet, name the first four columns **Test**, **Brand**, **Drywt**, and **Wetwt**. Figure 15.2 illustrates the variable names in the Untitled worksheet. Then input the data from Table 15.1 as these variables' values. *Remember to use numbers (1 and 2), not letters (A and B), for the variable Brand.*

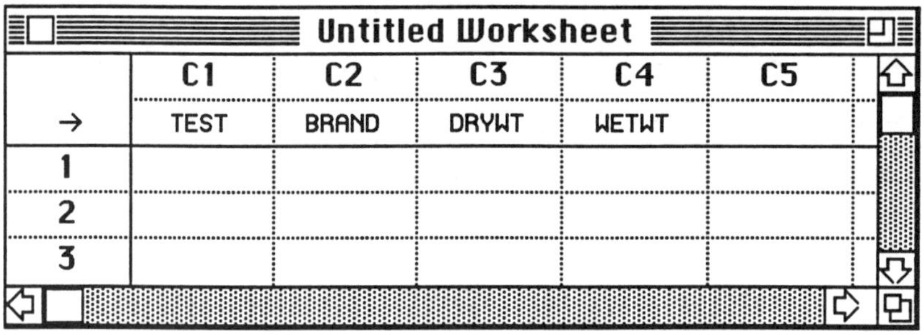

	C1	C2	C3	C4	C5
→	TEST	BRAND	DRYWT	WETWT	
1					
2					
3					

Figure 15.2 Untitled Worksheet

When you have finished typing the data and have double-checked it, save the data set onto your diskette by selecting **Save Worksheet As** under the **File** menu. Name your worksheet **Paper.Towels**.

We will now create a new variable, **Water**, to reflect the weight of the water absorbed by each paper towel. Water for each row (sheet) is the difference Wetwt - Drywt, and can be created as follows:

1. Under the **Calc** menu, click and hold on **Functions and Statistics** and then select **General Expressions** from the submenu. A General Expressions dialog box similar to Figure 15.3 will appear.

2. Click in the box to the right of **New/modified variable** and type **Water**.

3. Click in the box under **Expression** and type **'Wetwt'-'Drywt'**.

4. Click **OK**.

In a moment, the new variable, Water, should appear in your worksheet. The manner in which we have arranged the data, with the variable Water as the measure of interest and the variable Brand identifying the population from which the measurement came, is a standard way to arrange similar measurements from different groups. We will want to do graphics and descriptive statistics on Water separately

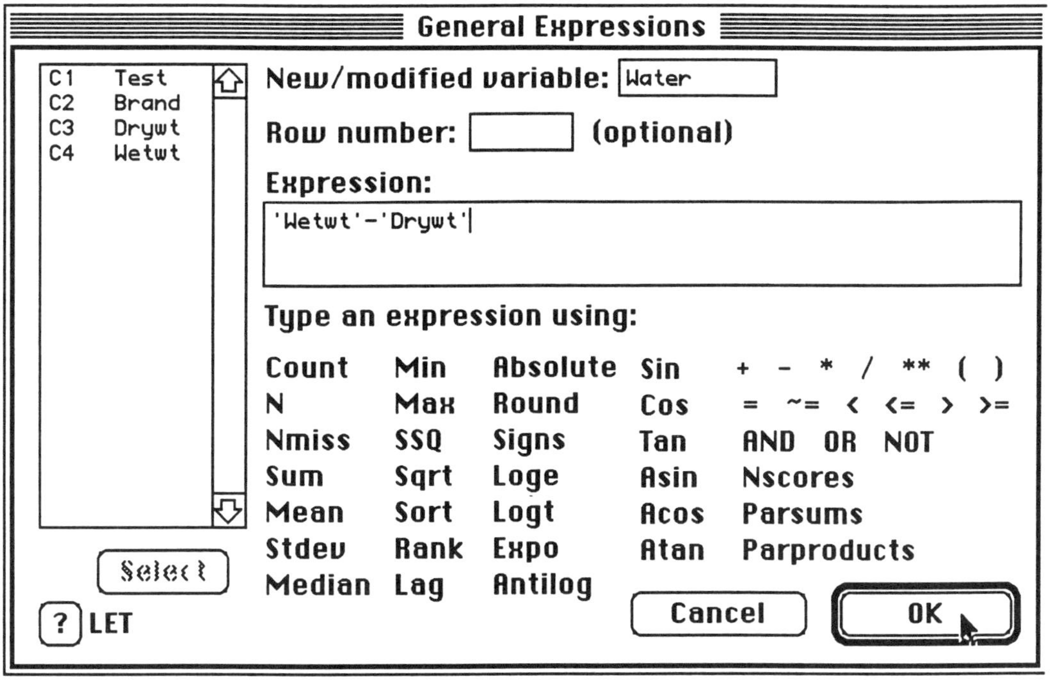

```
┌─────────────────────────────────────────────────────────────┐
│ ═══════════════════ General Expressions ═══════════════════ │
│ ┌──────────────┐                                            │
│ │ C1   Test  ▲ │  New/modified variable: │Water    │        │
│ │ C2   Brand   │                                            │
│ │ C3   Drywt   │  Row number: │       │  (optional)         │
│ │ C4   Wetwt   │                                            │
│ │              │  Expression:                               │
│ │              │  ┌──────────────────────────────────────┐  │
│ │              │  │'Wetwt'-'Drywt'│                      │  │
│ │              │  │                                      │  │
│ │              │  └──────────────────────────────────────┘  │
│ │              │                                            │
│ │              │  Type an expression using:                 │
│ │              ▼                                            │
│ │              │  Count   Min    Absolute  Sin   + - * / ** ( ) │
│ │              │  N       Max    Round     Cos   = ~= < <= > >= │
│ │              │  Nmiss   SSQ    Signs     Tan   AND  OR  NOT   │
│ │              │  Sum     Sqrt   Loge      Asin  Nscores        │
│ │   ┌───────┐  │  Mean    Sort   Logt      Acos  Parsums        │
│ │   │Select │  │  Stdev   Rank   Expo      Atan  Parproducts    │
│ │   └───────┘  │  Median  Lag    Antilog   ┌────────┐ ┌────┐    │
│ │ ┌─┐          │                           │ Cancel │ │ OK │    │
│ │ │?│ LET      │                           └────────┘ └────┘    │
└─────────────────────────────────────────────────────────────┘
```

Figure 15.3 General Expressions Dialog Box

for each Brand group. In this role, the population identifier Brand is sometimes called a **by-variable** or **subscript**.

We will begin by obtaining descriptive statistics for the Water variable for each brand:

1. Under the **Stat** menu, click and hold on **Basic Statistics** and then select **Descriptive Statistics** from the submenu. A Descriptive Statistics dialog box similar to Figure 15.4 will appear.

2. Click in the box under **Variables** and then double-click on **Water**.

3. Click in the box to the left of **By variable.**

4. Click in the box to the right of **By variable** and type '**Brand**'.

5. Click **OK**.

After a moment, descriptive statistics will appear in the Session window. Notice that there are two rows, one for each brand. Based on the two sample means, does one brand appear to absorb more water than the other?

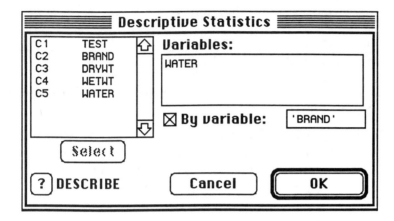

Figure 15.4 Descriptive Statistics Dialog Box

Compare the water weight absorbed for the two brands by constructing side-by-side boxplots:

1. Under the **Graph** menu, select **Boxplot**. A Boxplot dialog box similar to Figure 15.5 will appear.

2. Click in the box next to **Variable** and then double-click on **Water**.

3. Click in the box to the left of **By variable**.

4. Click in the box to the right of **By variable** and type **'Brand'**.

5. Click **OK**.

Before doing formal comparison of the two brands, print your boxplots by selecting **Print Window** under the **File** menu. Be careful not to pick up someone else's output. There may be several other students also using the printer. What differences do you find in the two boxplots?

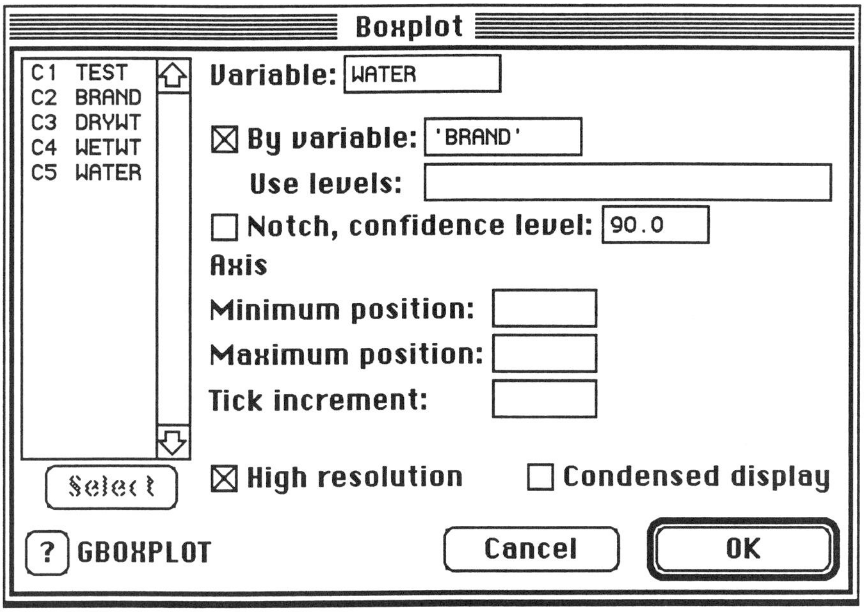

```
╔═══════════════════ Boxplot ═══════════════════╗
║ C1 TEST   ▲   Variable: [WATER]               ║
║ C2 BRAND                                      ║
║ C3 DRYWT      ☒ By variable: ['BRAND']        ║
║ C4 WETWT         Use levels: [        ]       ║
║ C5 WATER                                      ║
║               ☐ Notch, confidence level: [90.0]║
║               Axis                            ║
║               Minimum position: [  ]          ║
║               Maximum position: [  ]          ║
║               Tick increment:   [  ]          ║
║      ▼                                        ║
║  [Select]     ☒ High resolution   ☐ Condensed display ║
║                                               ║
║  [?] GBOXPLOT        [Cancel]   [  OK  ]      ║
╚═══════════════════════════════════════════════╝
```

Figure 15.5 Boxplot Dialog Box

To satisfy the assumptions of the pooled-variance two-sample t-test, the variability in the two samples should be roughly equal. Are the boxes in your boxplots of roughly equal length?

Finally, we will use Minitab to formally compare the population means, the mean weights of water that would have been absorbed had the entire rolls been tested. We will test the null hypothesis that μ_1, the mean weight of absorbed water if all of the name-brand roll were tested, is equal to μ_2, the mean weight of absorbed water for all of the store-brand roll. Our alternative hypothesis is that $\mu_1 > \mu_2$, since this would show that the store brand was incorrect in its claim of absorbency equal to that of the name brand. We will also obtain a 95 percent confidence interval for the difference in population means.

These analyses are based on the two-sample t statistic discussed in most introductory statistics texts. The one we will use is the one for independent samples (since we generated random numbers for each roll independently); it also assumes equal population variances. We can do both the hypothesis test and the confidence interval at the same time in Minitab:

1. Under the **Stat** menu, click and hold on **Basic Statistics** and then select **2-Sample t** from the submenu. A 2-Sample *t* dialog box similar to Figure 15.6 will appear.

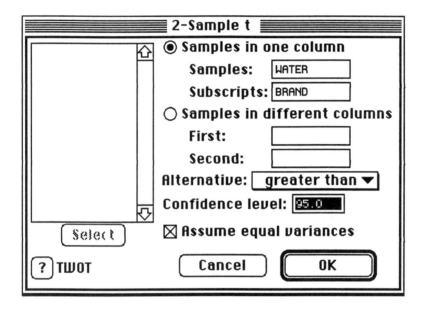

Figure 15.6 2-Sample t Dialog Box

2. Click in the circle to the left of **Samples in one column**.

3. Click in the box to the right of **Samples** and then double-click on **Water**.

4. Click in the box next to **Subscripts** and double-click on **Brand**.

5. To select the appropriate one-sided alternative, click and hold on the triangular button in the box next to **Alternative** and select **greater than** in the submenu.

6. The desired confidence level for the confidence interval is 95 percent, which is the default. If we had wanted a 90 percent interval we would have typed 90.0 in the box next to **Confidence level**.

7. Click in the box next to **Assume equal variances**.

8. Click **OK**.

The results of your two-sample *t* analyses will appear in the Session window. There is discussion on the interpretation of confidence intervals and hypothesis tests for differences in population means in Step 5 of Session 14. This may aid you in interpreting the results comparing these brands of paper towels. Are we able to conclude that the population mean for the name-brand towels is significantly larger than the population mean for the store-brand towels using a 5 percent significance level?

What information do you learn from the confidence interval for the difference in the population means?

Print your descriptive statistics, hypothesis test results, and confidence interval at this time by choosing **Print Window** under the **File** menu. Then quit Minitab by selecting **Quit** under the **File** menu. If you are working in a team on the computer, copy the data set **Paper.Towels** to each partner's diskette, as described in the "Copying Files Between Diskettes" section of Session 1.

PARTING GLANCES

When should we compare treatment/population means using the paired *t* statistic as opposed to the independent-samples *t* statistic? The paired *t* will give a sharper comparison if there is reason to believe that the measurements in a pair will vary together. In Session 14, two walking methods were compared in terms of walk time required to cover a certain distance. If a person can walk relatively quickly under one method, he or she should be relatively quick under the other method, too. There is reason to believe that these two measurements will vary together. The paired *t* experiment and data analysis was therefore a better choice than the independent-samples *t* in that case.

In this session, however, there is no natural pairing of paper towels of one roll with the other. Just because the third towel in the name-brand pile showed an unusually large weight of absorbed water, this is no reason to expect the third towel in the store-brand pile will be unusually high. Since there is no natural pairing, the better choice for comparison here is the independent-samples method.

Notice also that the design of the experiment and the choice of formal statistical inference method go hand in hand. Once we decided in the walking experiment to have each person walk both ways in turn, the option to use an independent-samples t-test analysis was gone. Never do an experiment without knowing pretty specifically what formal analyses will be done on the data. If you are not sure what design or analysis is appropriate, get help from a professional statistician *before* collecting data. Many colleges and universities have professional statisticians who help student and faculty researchers.

EXTENDED WRITING ASSIGNMENT

Refer to Appendix 1, "Technical Report Writing," and Appendix 2, "Technical Report Writing Checklist," for guidance on format and style for your report.

Write a report to the editorial staff of the consumer magazine mentioned under "The Setting." Your report should include:

1. A clear and complete description of the experiment and how the data was collected
2. Descriptive statistics, both numerical and graphical
3. A presentation and careful interpretation of the hypothesis test
4. A presentation and careful interpretation of the confidence interval

Give correct interpretations in language the editorial staff will understand.

Name   Section   Session 15

SHORT ANSWER WRITING ASSIGNMENT

All answers should be complete sentences. Include your boxplots, descriptive statistics output, and two-sample t output with this assignment.

1. What could cause the first five sheets on a single roll to be unrepresentative of the entire roll?

2. What could cause the sheets selected in a systematic random sample to be unrepresentative of the entire roll?

3. What do your boxplots tell you about the distribution of the amount of water absorbed for the two rolls?

4. Looking at the descriptive statistics and boxplots, do you think the assumption of equal population variances for the two-sample t-test is reasonable here? Why or why not?

5. Carefully interpret the results of your hypothesis test in language the consumer magazine editorial staff will understand.

6. Carefully interpret your confidence interval for the difference in population means in language the consumer magazine editorial staff will understand.

SESSION SIXTEEN

Variation in Nature

INTRODUCTION

Investigators frequently make several different measurements on an individual sample item. In this experiment we will be making four different measurements on hickory nuts. Each measurement will reflect a different aspect of the size of the nut. For large nuts, each of the measures will be relatively large. For small nuts, each of the measures will be relatively small. It follows from this logic that the correlation between any two of these measurement methods will be greater than zero.

In some measurement problems, it is relatively easy to make some of the measurements and relatively difficult to make another measurement. In such settings, statisticians use a technique known as **regression** to predict the value of the difficult measurement based on the values of the relatively easy measurements.

STATISTICAL CONCEPTS

Scatter plots, simple linear regression, correlation.

MATERIALS NEEDED

For each team, 27 hickory (or a local variety) nuts, a dial caliper capable of measuring to the nearest 0.01 inch, a balance sensitive to the nearest gram, 2 egg cartons with positions numbered from 1 to 24, and 2 objects with premeasured dimensions (less than 6 inches) and premeasured weight.

MEASURING OUTSIDE DIAMETERS WITH A DIAL CALIPER

Place the object between the lower jaws of the caliper. Close the jaws on the object to get a snug fit. The measurement is the sum of two components. The numbers on the bottom of the body of the caliper represent inches. The outside numbers on the dial are in units of 0.01 inches. The measurement depicted in Figure 16.1 represents $1.00 + 0.70 = 1.70$ inches.

Your instructor will provide some objects with known dimensions. Practice measuring these objects before you begin the experiment.

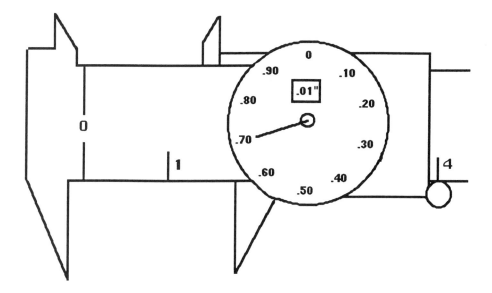

Figure 16.1 Dial Caliper

WEIGHING WITH A DIGITAL ELECTRONIC BALANCE

Most balances can weigh in either grams or ounces. Set your balance to weigh in grams if possible. Make sure the balance is on a level surface. Turn on the balance. When zero is displayed, the balance is ready for use. Place the object to be weighed on the platform of the balance, and read the displayed weight.

THE SETTING

You are a research assistant for a nutty botany professor. For a number of years, he has been interested in a particular species of hickory tree. In the current research, the research team wishes to study the size of hickory nuts and use this information to predict the weight of the nut. The work of making the measurements has been left to you.

BACKGROUND

You have recently walked through a forest and collected a sample of 24 hickory nuts for use in your study.

THE EXPERIMENT

STEP 1: OPERATIONAL DEFINITIONS

Hickory nuts are irregularly shaped three dimensional objects. We will define the **top** of the nut to be the point where it was attached to the tree branch. Note that the pieces of the shell come together at the top. The **bottom** of the nut is the point opposite the top where the pieces of the shell come together. Define the **height** of the nut to be the distance from the top to the bottom. Holding the nut with the top up, rotate the nut until the side-to-side distance is maximized. This maximum side-to-side distance will be called the **major axis**. Now, rotate the nut until the side-to-side distance is minimized. This minimum side-to-side distance will be called the **minor axis**. Figure 16.2 illustrates these dimensions.

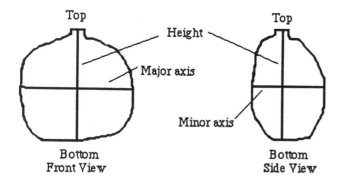

Figure 16.2 Hickory Nut

STEP 2: DATA COLLECTION

Your research team has been given a sample of 24 hickory nuts. You will be measuring (1) the height, major axis, and minor axis of the nuts to the nearest .01 inch with the caliper and (2) the weight of the nuts to the nearest gram using the electronic balance. We are interested in the relationships among the four measurements. *Make and record all four measurements on one nut and place the nut in the egg carton location corresponding to its number before going to the next nut.* Record the measurements in Table 16.1.

STEP 3: DATA ANALYSIS

Launch Minitab and insert a diskette. Before entering your data, name the first four columns **Height**, **MajAxis**, **MinAxis**, and **Weight**. The worksheet should now look like Figure 16.3. Carefully enter the data from Table 16.1 into the appropriate columns of the worksheet. Double-check that the data has been entered correctly. When you are sure that your data is correct, save it on your diskette as **Nutdata** by selecting **Save Worksheet As** from the **File** menu.

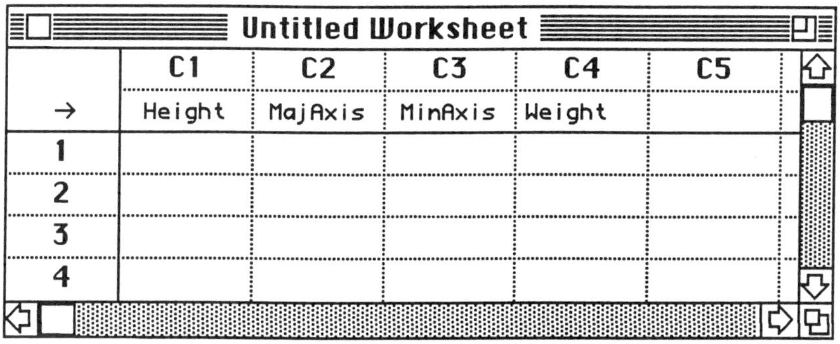

	C1	C2	C3	C4	C5
→	Height	MajAxis	MinAxis	Weight	
1					
2					
3					
4					

Figure 16.3 Untitled Worksheet

Let us first look at descriptive statistics for each of the four variables.

1. Under the **Stat** menu, click and hold on **Basic Statistics** and then select **Descriptive Statistics** in the submenu. A Descriptive Statistics dialog box similar to Figure 16.4 will appear.

Table 16.1 Hickory Nut Data

Nut	Height	Major Axis	Minor Axis	Weight
1				
2				
3				
4				
5				
6				
7				
8				
9				
10				
11				
12				
13				
14				
15				
16				
17				
18				
19				
20				
21				
22				
23				
24				

Table 16.2 Descriptive Statistics

Variable	Minimum	Maximum	Sample Mean	Sample Median	Standard Deviation
Height					
Major axis					
Minor axis					
Weight					

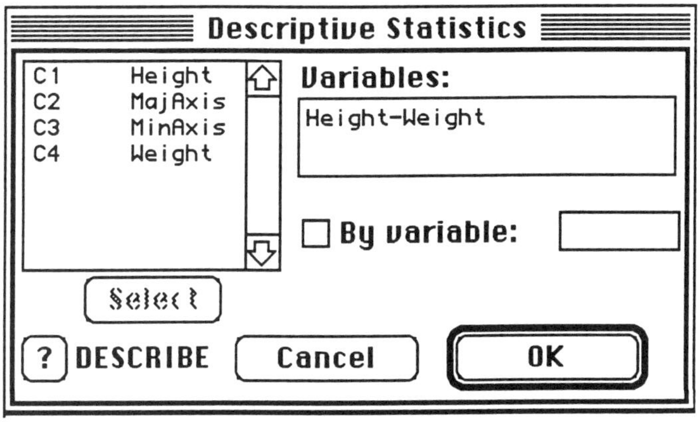

Figure 16.4 Descriptive Statistics Dialog Box

2. Click in the box under **Variables** and then double-click on each of **Height**, **MajAxis**, **MinAxis**, and **Weight**.

3. Click **OK**.

A set of descriptive statistics for each variable will appear in the Session window. Look at the minimum, maximum, and mean values for each variable. Are these values reasonable? If not, you have probably made a data-entry error. If you suspect an error, look at the worksheet again by selecting **Data** under the **Window** menu. If you find any errors, correct the data and repeat the computation of the descriptive statistics. After you are sure that your data is correct, complete Table 16.2.

278 SESSION SIXTEEN

Notice that all four of our variables reflect some aspect of the size of the nut. Do you expect that large values of one variable will tend to occur with large values of the other variables and that small values of one variable will tend to occur with small values of the other variables?

We are now ready to look at the relationships between the variables. We will do this graphically by producing scatter plots and numerically by computing sample correlation coefficients. Scatter plots are graphical presentations of data for two variables. Because we have four variables, there are six different pairs of variables for our experiment.

Before producing each scatter plot, we need to decide which variable will be displayed on the vertical axis and which variable will be displayed on the horizontal axis. For example, to produce a scatter plot with Weight on the vertical axis and Majaxis on the horizontal axis, we do the following:

1. Under the **Graph** menu, select **Scatter Plot**. A Scatter Plot dialog box similar to Figure 16.5 will appear.

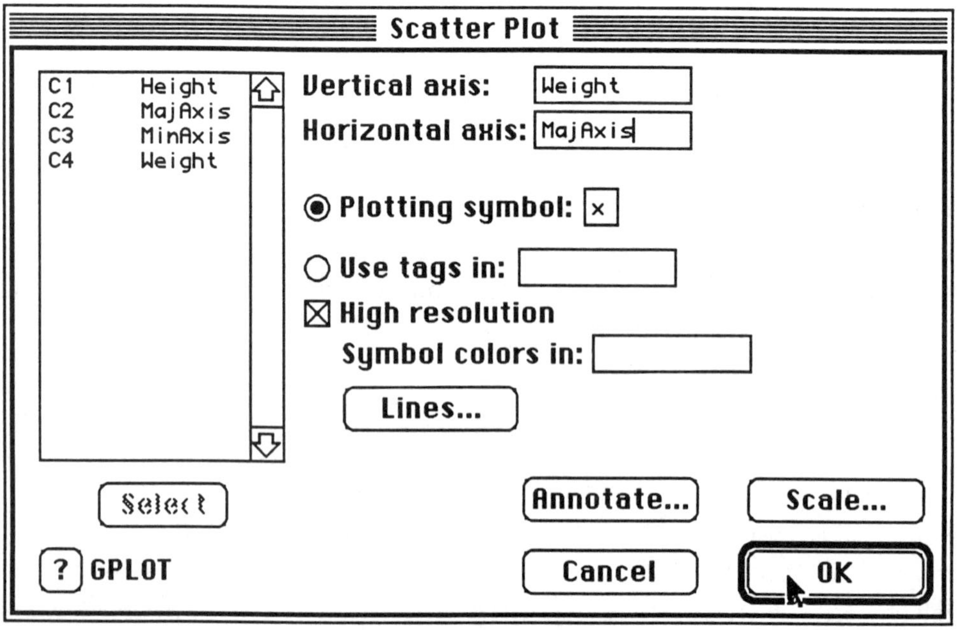

Figure 16.5 Scatter Plot Dialog Box

2. Click in the box to the right of **Vertical axis** and type **Weight**.

3. Click in the box to the right of **Horizontal axis** and type **MajAxis**.

4. Click **Annotate**. An Annotate Scatter Plot dialog box similar to Figure 16.6 will appear.

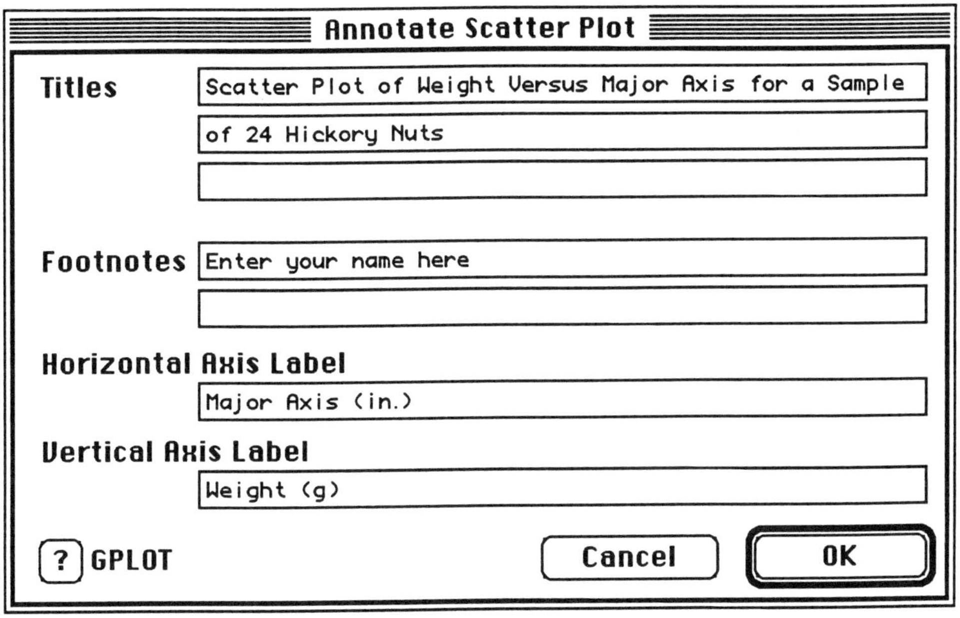

```
┌─────────────── Annotate Scatter Plot ───────────────┐
│ Titles     [Scatter Plot of Weight Versus Major Axis for a Sample]
│            [of 24 Hickory Nuts                                    ]
│            [                                                      ]
│
│ Footnotes  [Enter your name here                                  ]
│            [                                                      ]
│
│ Horizontal Axis Label
│            [Major Axis (in.)                                      ]
│ Vertical Axis Label
│            [Weight (g)                                            ]
│
│ [?] GPLOT            [ Cancel ]    [   OK   ]
└──────────────────────────────────────────────────────┘
```

Figure 16.6 Annotate Scatter Plot Dialog Box

5. Click in the box to the right of **Titles** and type an appropriate title.

6. Click in the box to the right of **Footnotes** and type your name. This will allow you to identify your scatter plot when it comes off the printer.

7. Click in the box below **Horizontal Axis Label** and type **Major Axis (in.)**.

8. Click in the box below **Vertical Axis Label** and type **Weight (g)**.

9. Click **OK**. The Scatter Plot dialog box will reappear.

10. Click **OK**.

The scatter plot will now appear on the screen. To get a printout of your plot, select **Print Window** under the **File** menu. If you are sharing a computer, make enough copies for all partners.

As you look at the scatter plot, ask yourself the following questions:

1. Does there appear to be a relationship between the two variables? We anticipated that large values of each variable would occur together and that small values would occur together.

2. If there is a relationship, do the points tend to lie along a straight line?

3. Do the points fall exactly on a line or smooth curve, or is there variation about the line or curve?

Summarize your thoughts about this scatter plot in the top cell of Table 16.3.

Produce additional scatter plots and write appropriate comments in Table 16.3 for each of the remaining pairs of variables (as printouts will not be necessary, you can omit items 4–9 in the above scatter plot instructions):

1. Height versus Majaxis
2. Height versus Minaxis
3. Height versus Weight
4. Majaxis versus Minaxis
5. Weight versus Minaxis

Statisticians quantify the strength of the linear relationship between two variables by the **sample correlation coefficient**. The sample correlation coefficient is given by

$$r = \frac{\Sigma[(x - \bar{x})(y - \bar{y})]}{[\Sigma(x - \bar{x})^2 \Sigma(y - \bar{y})^2]^{1/2}}$$

where x and y represent the two variables and the summation is over all observations. The value of r is always between -1 and 1. A perfect linear relationship with positive slope has a correlation of 1 and a perfect linear relationship with negative slope has a correlation of -1. A correlation near zero suggests little or no linear relationship between the variables.

VARIATION IN NATURE 281

Table 16.3 Comments on Scatter Plots

Weight Versus Major Axis
Height Versus Major Axis
Height Versus Minor Axis
Height Versus Weight
Major Axis Versus Minor Axis
Weight Versus Minor Axis

Table 16.4 Sample Correlation Coefficients

Variables	Correlation	Variables	Correlation
Height, Major Axis		Height, Minor Axis	
Height, Weight		Major Axis, Minor Axis	
Weight, Major Axis		Weight, Minor Axis	

To compute correlation statistics in Minitab:

1. Under the **Stat** menu, click and hold on **Basic Statistics** and then select **Correlation** in the submenu. A Correlation dialog box similar to Figure 16.7 will appear.

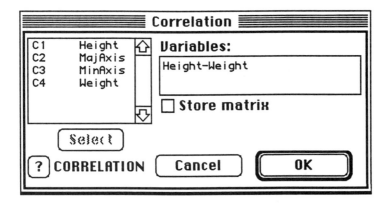

Figure 16.7 Correlation Dialog Box

2. Click in the box under **Variables** and double-click on each of **Height**, **MajAxis**, **MinAxis**, and **Weight**.

3. Click **OK**.

The sample correlation coefficients for all pairs of variables will now appear in the Session window. Write the sample correlation coefficients in Table 16.4.

Based on the correlation coefficients, which pair of variables has the strongest linear relationship? Does this agree with what you saw in the scatter plots?

Let's now develop a method for predicting the weight of a nut based on its external measurements. Using one or more variables to predict the value of another variable is an example of **regression**. In regression, one tries to predict the response (dependent) variable, Weight, using some function of Height, Majaxis, and Minaxis. To keep things simple, let's use the single predictor (independent) variable Majaxis and the **simple linear regression** model,

$$\text{Weight} = \beta_0 + \beta_1(\text{Majaxis}) + \text{error},$$

where β_0 is the intercept of the line, β_1 is the slope of the line, and error reflects random variation. The error term reflects the fact that all nuts with the same Majaxis measurement do not weigh the same amount. For this model to be reasonable, we need the scatter plot of Weight versus Majaxis to show a pattern that can be approximated by a straight line. Does your plot of Weight versus Majaxis show such a pattern?

We will now use Minitab to fit a regression line to the data. To do this, we need to estimate β_0 and β_1, the intercept and the slope of the model. We will use a method known as **least squares** to estimate β_0 and β_1. To fit a regression line by least squares in Minitab:

1. Under the **Stat** menu, click and hold on **Regression** and then select **Regression** in the submenu. A Regression dialog box similar to Figure 16.8 will appear.
2. Click in the box to the right of **Response** and type **Weight**.
3. Click in the box to the right of **Predictors** and type **MajAxis**.
4. Click **OK**.

The results of the regression analysis will appear in the Session window. An example of a portion of the results for a similar data set appears in Table 16.5.

The estimates of β_0 and β_1 appear in the **Coef.** column. For the example data, the estimate of β_0 is -15.214 and the estimate of β_1 is 25.989. The **R-sq.** entry is the value of the R-square statistic. This gives the percentage of variation in Weight that is explained by the regression model. For simple linear regression, R-square is the square of the correlation statistic r. For the example, 47.3 percent of the variation in

Table 16.5 Minitab Regression Results for Example Data

The regression equation is
Weight = - 15.2 + 26.0 MajAxis

Predictor	Coef	Stdev	t-ratio	p
Constant	-15.214	6.397	-2.38	0.026
MajAxis	25.989	5.725	4.54	0.000

s = 1.320   R-sq = 47.3%   R-sq(adj) = 45.0%

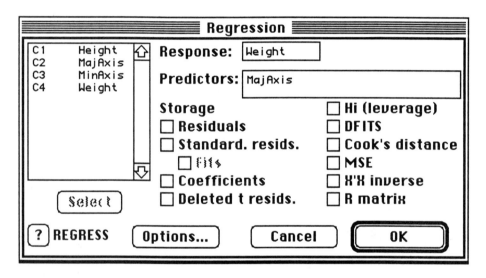

Figure 16.8 Regression Dialog Box

Weight was explained by the model. The sample regression line for the example data is

Weight = -15.214 + 25.989(Majaxis).

To predict the Weight of a nut based on its Majaxis, we plug its Majaxis measurement into the equation of the sample regression line. For the example data, a nut with

Table 16.6 Results of Regression of Weight on Major Axis

Parameter	Estimate
Intercept, β_0	
Slope, β_1	

Sample regression line

Predicted weight = _____ + _____ Majaxis

R-squared = _____ %

Majaxis of 1.00 inch would have a predicted weight of

$$\text{Weight} = -15.214 + 25.989\,(1.00) = 10.775 \text{ grams}$$

Use the information from the Session window to complete Table 16.6 for your data.

The product of Height, Majaxis, and Minaxis is roughly proportional to the volume of the nut. As this product takes three dimensions into account, it may be a better predictor of Weight than the single dimension Majaxis. To create the new variable **Product**:

1. Under the **Calc** menu, click and hold on **Functions and Statistics** and then select **General Expressions** in the submenu. A General Expressions dialog box similar to Figure 16.9 will appear.
2. Click in the box to the right of **New/modified variable** and type **Product**.
3. Click in the box below **Expression** and type **'Height' * 'MajAxis' * 'MinAxis'**.
4. Click **OK**.

The variable Product now appears in the worksheet in the next available column, C5. Now, produce and print a scatter plot of Weight versus Product. Is a straight-line model reasonable for predicting the variable Weight based on the variable Product?

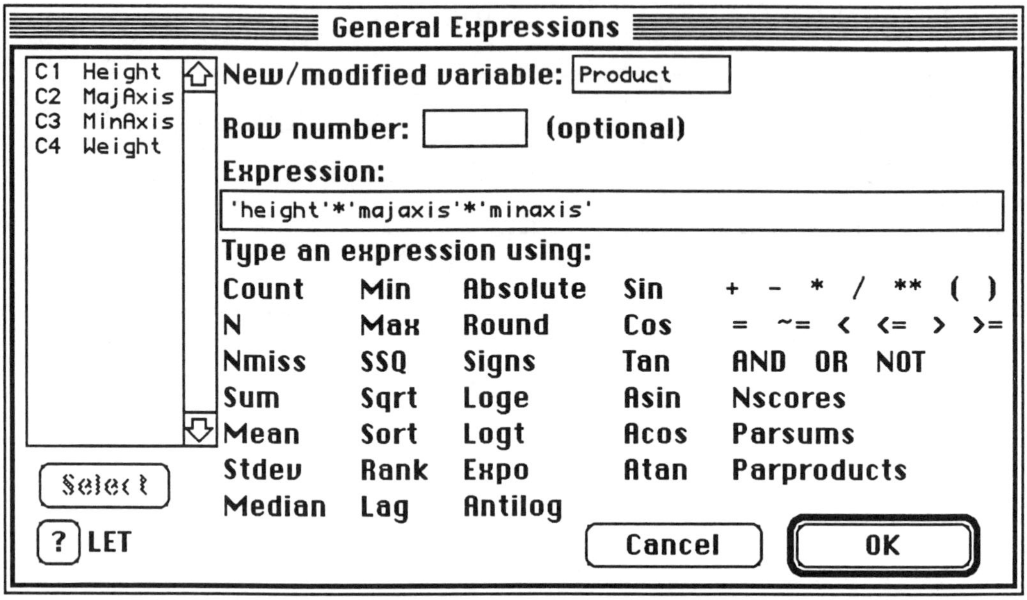

```
┌─────────────────────────────────────────────────────────────┐
│                    General Expressions                      │
│ C1  Height  ⇧ New/modified variable: [Product]              │
│ C2  MajAxis                                                 │
│ C3  MinAxis   Row number: [      ] (optional)               │
│ C4  Weight   Expression:                                    │
│              ['height'*'majaxis'*'minaxis'               ]  │
│              Type an expression using:                      │
│              Count    Min    Absolute   Sin   + - * / ** ( )│
│              N        Max    Round      Cos   = ~= < <= > >=│
│              Nmiss    SSQ    Signs      Tan   AND  OR  NOT  │
│              Sum      Sqrt   Loge       Asin  Nscores       │
│            ⇩ Mean     Sort   Logt       Acos  Parsums       │
│  [Select]    Stdev    Rank   Expo       Atan  Parproducts   │
│  [?] LET     Median   Lag    Antilog                        │
│                                    [Cancel]  [   OK   ]     │
└─────────────────────────────────────────────────────────────┘
```

Figure 16.9 General Expressions Dialog Box

Fit the regression of the response variable Weight on the predictor variable Product, and put the results in Table 16.7. Follow the Minitab directions you used in fitting the other regression line but use **Product** in place of MajAxis in item 3.

We compare two simple linear regression models by comparing their R-square values. Which predictor variable, Majaxis or Product, yields the higher R-square value; that is, which regression explains more of the variation in the response variable Weight?

Ask your instructor for three additional hickory nuts. We want to use the regression equation with the predictor variable Product to predict the Weight of the additional nuts. Measure the Height, Majaxis, and Minaxis with the caliper and compute the value of Product for these three nuts. Write the results in Table 16.8.

Evaluate the regression equation in Table 16.7 using the value of Product for each additional nut. The resulting values are the regression estimates of the Weight for the new nuts. Write these estimates in Table 16.9.

Table 16.7 Results of Regression of Weight on Product

Parameter	Estimate
Intercept, β_0	
Slope, β_1	

Sample regression line
Predicted weight = _____ + _____ Product
R-squared = _____ %

Table 16.8 External Measurements on Additional Nuts

Additional Nut	Height	Major Axis	Minor Axis	Product
1				
2				
3				

Table 16.9 Estimated Weights and Actual Weights of Additional Nuts

Additional Nut	Estimated Weight	Actual Weight
1		
2		
3		

It is now the moment of truth! Weigh the additional nuts and write their actual weights in Table 16.9. Did the regression model do a good job of predicting the weights of these nuts?

This concludes our calculations on the data. Make sure you have all the output you need, and then quit Minitab by selecting **Quit** from the **File** menu.

PARTING GLANCES

We have demonstrated the usefulness of regression by predicting the weight of a nut. In many important applications, determining the true value of the response variable can be costly, time-consuming, or impossible. Consider the problem of deciding whether to admit a student to college. Admission officials often wish to predict the student's first-year grade-point ratio to help them make the admission decision. Clearly this true grade-point ratio is not available at the time the admission decision has to be made. Often scores from standardized tests and measures of high school rank are used as predictor variables in a regression model to estimate what the student's first-year grade-point ratio would be.

In our hickory nut example, we used one predictor variable, either major axis or product, in the regression model. In some applications more than one predictor variable is used. A regression model with more than one predictor is called a **multiple regression** model. In predicting first-year grade-point ratios, we might use three predictor variables: verbal SAT score, quantitative SAT score, and relative rank = rank in high school class/size of high school class. A possible regression model would be

Grade-Point Ratio =
$\beta_0 + \beta_1$(Verbal) + β_2(Quantitative) + β_3(Relative Rank) + error

In other regression models we **transform** the data before fitting the regression model. For example, we might try to predict the logarithm of the weight of the hickory nut using a regression model where the predictor variables were the logarithms of the height, the major axis, and the minor axis.

EXTENDED WRITING ASSIGNMENT

Refer to Appendix 1, "Technical Report Writing," and Appendix 2, "Technical Report Writing Checklist," for guidance on format and style for your report.

Scientists summarize the results of experiments in articles printed in scientific publications known as journals. Write a report that summarizes your experiment. It should include:

1. A description of how the sample was collected

2. A description of the experiment you performed including operational definitions for the variables you measured

3. Graphical and numerical measures of the relationships among the four measurement variables and an explanation of what these measures mean

4. A proposed method for estimating the weight of a hickory nut based on its external measurements

5. Any recommendations for further experimentation based on your results.

Name\tSection\tSession 16

SHORT ANSWER WRITING ASSIGNMENT

All answers should be complete sentences. Include copies of your scatter plot of Weight versus Product and Tables 16.7, 16.8 and 16.9 with this assignment.

1. Based on the data in Table 16.4, what was the correlation between the variables Weight and Height?

2. Explain in nontechnical terms what this correlation statistic implies about the joint behavior of these two variables.

3. Are the values of the correlation statistics consistent with the statement "Hickory nuts grow simultaneously in all dimensions"? Explain.

4. Does your scatter plot of Weight versus Product show that the points tend to fall along a straight line with positive slope? If not, what pattern do you see in the scatter plot?

5. We used two different predictor variables, Majaxis and Product, in our regression models for Weight. Which predictor (independent) variable do you prefer? Why?

6. Show how you used the information from Tables 16.7 and 16.8 to compute the estimated weight for the first additional nut.

SESSION SEVENTEEN
Random Sampling

INTRODUCTION

How many times have you heard on the news or read in the headlines a catchy phrase such as, "A sample of eligible voters in the United States showed 89 percent in favor of Candidate Simsick." Clearly this reported statistic makes us think Candidate Simsick is a sure thing. But several questions regarding the number "89 percent" should be considered before it can be perceived as a believable statistic.

In almost all applications of statistics, the first step is to obtain a random sample. The importance of selecting the sample in a statistically correct manner cannot be overemphasized. If the random sample is not representative of the population, any statistics and conclusions based on the sample are inappropriate for the larger population.

In the above example, the population of interest would be "all eligible voters in the United States." It is not feasible to poll everyone, so a sample is taken to learn about the percentage who favor Candidate Simsick. How do you think this random sample of voters was obtained? Perhaps the pollster stood in front of the Republican headquarters of some small town. Can you see the flaw in this sampling scheme? Another important aspect of a sample is its size. The percentage, 89 percent, could have been based on a sample of size 18 (16 favoring Candidate Simsick) or one of size 1000 (890 favoring). Larger samples are preferable, since the sample percentage will tend to be closer to the population percentage.

STATISTICAL CONCEPTS

Random sample, population, sampling frame, proportion, variability of a sample statistic, confidence interval for a proportion.

MATERIALS NEEDED

For each student, a calculator.

THE SETTING

You are the marketing representative for a local car dealership. By selecting a random sample, you wish to learn about certain characteristics of a population of cars and trucks.

BACKGROUND

During the 1970s and 1980s, sales of Japanese-made cars and trucks skyrocketed, while domestic sales were sluggish. Generally, the public believed foreign cars and trucks were better made and more reliable. This attitude, perhaps true at the time, was due in part to statistical quality control methods implemented by the Japanese during the 1950s and 1960s.

In the 1980s, U.S. automobile manufacturers began energetically practicing the principles of total quality management. Their hard work and new management philosophy seem to be paying off. Most people would agree that domestic cars and trucks are better products today than they were 15 years ago. Recent trends in consumer purchasing show that Americans are beginning to see the improved quality of domestic vehicles and that sales are increasing.

Based on a sample of vehicles from a nearby parking lot, you are to estimate the proportion of Japanese-made cars and trucks in a particular population.

THE EXPERIMENT

STEP 1: THE LOCATION

You should choose the location to be a large, nearby parking lot. The cars and trucks in the parking lot will serve as the population. It is important to keep in mind that sampling from this parking lot allows us to learn about the population of *cars and trucks in that lot*. This may or may not be representative of the entire population of cars and trucks in your city. Think about the particular lot you will be using and whether the vehicles are representative of what the general public drives. Describe the location of your chosen lot and identify the population of interest.

STEP 2: RANDOM SAMPLES AND SAMPLING FRAMES

Let us begin with a formal definition of a "simple random sample." For our purposes, a **sample** is said to be **random** if it is selected in such a way that each member of the population is equally likely to be chosen and the members of the sample are chosen independently of one another.

In this setting, the first condition means that each vehicle should have an equal chance of being selected. The second requirement implies that one vehicle selection should not affect a future selection. We will use random numbers to ensure that both requirements are satisfied.

Before a sample can be taken, a **sampling frame** must first be defined. A sampling frame is an organized "list" of all the sampling units in the population. It is what we will eventually draw from to get the random sample. Each item in the list is assigned a unique number.

Here, our list will be a sketch of the parking lot, with each space numbered. Figure 17.1 contains the sampling frame for an example rectangular parking lot.

1	2	3	4	5	...	36	37	38	39	40
41	42	43	44	45	...	76	77	78	79	80

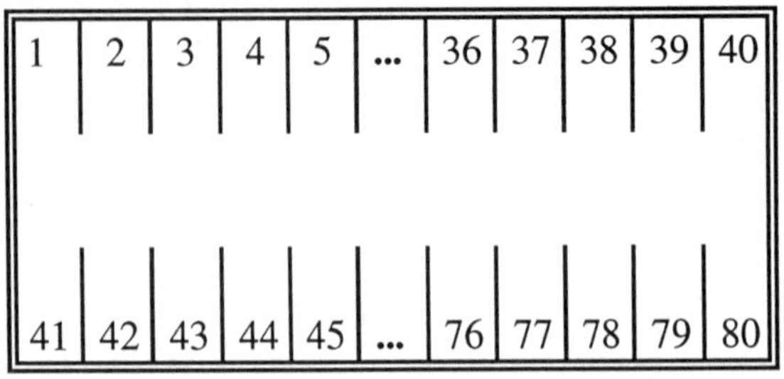

Figure 17.1 Sampling Frame for Example Rectangular Parking Lot

Obtain an illustration of the parking lot, making sure it is an accurate depiction. Number the spaces in some organized manner as in the rectangular lot example. In the text that follows, N denotes the total number of spaces.

STEP 3: RANDOM NUMBER GENERATION

Before you go out to the parking lot, you must first decide which cars and trucks will be chosen. To make each vehicle have the same chance of being chosen, we shall use the sampling frame from Step 2 and a list of random numbers that are generated using Minitab. The random numbers will tell us which parking-lot spaces will be selected in the sample. For example, if 45 comes in the list of generated numbers, then the vehicle parked in space 45 will be chosen in our sample. We need numbers between 1 and N.

Launch Minitab. To generate random integers using Minitab:

1. Under the **Calc** menu, click and hold on **Random Data** and then select **Integer** from the submenu. An Integer Distribution dialog box similar to Figure 17.2 will appear.

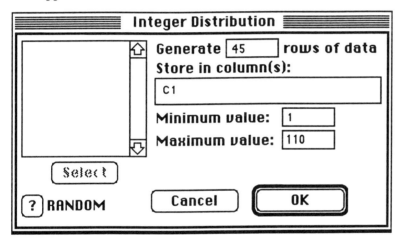

Figure 17.2 Integer Distribution Dialog Box

2. Click in the box to the right of **Generate** and type the number of random integers that you desire. For this experiment, 45 should be sufficient.

3. Click in the box beneath **Store in column(s)** and type **C1**.

4. Click in the box to the right of **Minimum value** and type **1**.

5. Click in the box to the right of **Maximum value** and type the number of spaces in your parking lot. If your lot has $N = 110$ spaces, then type 110.

6. Click **OK**.

After a short delay, random integers between 1 and *N* will appear in column C1 of the worksheet. We now want to get a hard copy of these random numbers. Move to the Session window by selecting **Session** under the **Window** menu.

In the Session window, click to the right of **MTB>** and type **Print C1**. Figure 17.3 illustrates the Session window. The random integers from C1 will now be listed in the Session window. To get a hard copy, select **Print Window** under the **File** menu. You will need this printout when you go to the parking lot. After you get your printout, quit Minitab by selecting **Quit** under the **File** menu.

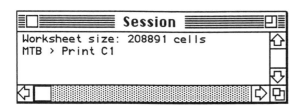

Figure 17.3 Session Window

STEP 4: COLLECTING THE DATA

Before the data is collected, we must be more precise about what the term Japanese-made means. Technically speaking, an "operational definition" for Japanese-made is needed so others can understand what we are estimating. Clearly there are many different ways to define the term *Japanese-made*. For example, many models that originated in Japan are now manufactured in the United States. What we will do here is use the three largest Japanese automobile companies: Honda, Toyota, and Nissan. Therefore, for each selected car or truck you should note whether it is made by one of Honda, Toyota, or Nissan.

Think of a **second characteristic** about cars that interests you. Examples are the proportion of illegal parkers, the proportion of cars and trucks with a certain brand of tires, or the proportion of cars and trucks with more than one bumper sticker. Describe the additional attribute that you will record about each vehicle.

$$SE(\hat{p}) = \sqrt{\frac{\hat{p}(1-\hat{p})}{n}}.$$

Fill in Table 17.2 with your data and the calculated values of $\hat{p}$ and $SE(\hat{p})$.

So far, you have calculated what are commonly called **point estimates**. You can say something like, "I estimate the proportion of cars and trucks that are red to be 0.34." Unfortunately, it is extremely unlikely that 0.34 is the population proportion. **Confidence intervals** are useful because they provide us with an interval that has a known chance of containing the true population proportion. The general formula for a confidence interval for a population proportion is

$$\hat{p} \pm z_{\alpha/2} \sqrt{\frac{\hat{p}(1-\hat{p})}{n}},$$

where $z_{\alpha/2}$ is the $100(1-\alpha/2)$ percentile from the standard normal distribution. For a 95 percent confidence interval, $\alpha = .05$ and $z_{\alpha/2} = 1.96$; for a 99 percent confidence interval, $\alpha = .01$ and $z_{\alpha/2} = 2.576$. The formula works best when n is large, preferably larger than 30. We shall use it here, with our moderate-size sample, but a larger sample would have been preferable. Complete Table 17.2 by calculating the confidence intervals for p_1 and p_2.

PARTING GLANCES

One should ask two important questions before selecting a sample: What is the population of interest? Is the sampling frame representative of this population?

Once a sample is collected, the numerical calculations necessary for estimating a proportion are relatively simple. But because it is based only on a sample and not the entire population, the point estimate will seldom be exactly equal to the true value. A confidence interval provides an interval that we can feel reasonably confident contains the true proportion.

We have also seen that the actual selection of individual units of the sample should be done randomly and objectively, which usually involves generated random numbers. When the units are not randomly selected, this introduces what is called sampling bias. Unfortunately, bias in sampling occurs too often. A historical example is the 1936 U.S. presidential election. The candidates were the incumbent, Franklin D. Roosevelt, and Governor Alfred Landon of Kansas. The country was recovering from the Great Depression, with millions still unemployed. A very prestigious publication called the *Literary Digest* conducted a poll based on 2.4 million individuals and predicted that Landon would win with 57 percent of the vote. The magazine had called the winner in every presidential election since 1916. But Roosevelt won the election in a big way, with 62 percent of the vote. What went wrong? It turns out that the *Literary Digest* picked their sample by using sources such as telephone directories and club membership lists. While the sample was chosen randomly, it was the sampling frame that was incorrect. Do you see why? By considering only those who owned telephones or were members of clubs, the pollsters were screening out the less well-to-do voters, who in the days of the depression constituted a large percentage of the voters. In addition, only 20 percent of the people who were polled responded. This caused a second source of bias called **nonresponse bias**.

REFERENCES

Bryson, M. C. (1976), "The Literary Digest Poll: Making of a Statistical Myth," *The American Statistician* 30, 184–185.

Freedman, D., Pisani, R., Purves, R., and Adhikari, A. (1991), *Statistics,* 2nd ed. (New York: Norton).

EXTENDED WRITING ASSIGNMENT

Refer to Appendix 1, "Technical Report Writing," and Appendix 2, "Technical Report Writing Checklist," for guidance on format and style for your report.

Write a report summarizing your findings. You should address the following issues:

1. A clear description of the population and the sampling frame

2. The manner in which the sample was obtained

3. The data, the estimates, the intervals, and careful interpretations

4. Suggestions for improved sampling procedures

Name_____ Section_____ Session 17

SHORT ANSWER WRITING ASSIGNMENT

All answers should be complete sentences. Include a copy of Table 17.1 with this assignment.

1. Briefly describe the location of the parking lot.

2. Are the cars and trucks representative of what the general public drives? If you believe so, state why you hold this view. If not, what group of drivers does your lot represent?

3. Briefly describe how Minitab was used to select the sample of cars and trucks.

4. Discuss how our sampling scheme satisfies both parts of the definition of a random sample.

5. Calculate a 99 percent confidence interval for p_1. How does it compare to the 95 percent interval?

6. What is $\hat{p}_2 = X_2/n$, and what does it estimate?

7. A **convenience sample** is one in which the individual units are chosen because they are convenient or easy to get. It is an incorrect way of drawing a sample that often leads to biased, misleading results. In this experiment, a convenience sample could have been taken by selecting the first 25 cars that we saw in the parking lot. Briefly discuss why this convenience sample might be biased.

APPENDIX ONE

Technical Report Writing

INTRODUCTION

Clear, honest, and efficient communication of results is absolutely vital to any research effort. The information you've obtained will be of little use if it is not effectively shared with others. Employers consistently report that communication skills are one of three common deficiencies they see in most technical-major college graduates (the other two are teamwork and statistical skills). This appendix is a set of simple guidelines to help with report writing for the extended writing assignments of the *Elementary Statistics Laboratory Manual*. Do not stop here in developing communication skills. Put genuine enthusiasm and energy into continued improvement in both written and oral communication. Take courses in technical writing and public speaking.

From the start, understand that technical report writing is very different in its goals from other writing tasks. The goals of report writing are stated in the first sentence of this appendix: clear, honest, and efficient communication of results. If the report is also pleasant to read, that's wonderful, but it is secondary. Moreover, what makes a report pleasant to read may surprise you: It should be clearly and logically organized, be thorough but not verbose, have neat and completely labeled tables and graphs, and so on. The person reading the report is not doing it for his or her entertainment. It's work. If the report is not clear, if unanswered questions come to the reader's mind, it is hard and annoying work. The report will be pleasant to read if it is not hard work to read it.

The report-writing process will vary according to individual tastes, but usually begins with rough outlines, followed by one or more rough drafts, leading to the final report. It is not a good idea to try to accomplish all of this in one sitting; allow some time between refinements for ideas to develop. Read this appendix carefully before you begin, and refer to Appendix 2, "Technical Report Writing Checklist," after each outline or draft refinement. Your instructor also may provide you with a scoring rubric, a form used in grading your reports. The rubric shows points allocated to various aspects of the report. If a rubric is provided, you should also refer to it carefully before you begin writing, and as a check of each draft of your report.

In all communication you should strive to know your audience in advance. What do they need from you? What is their background? Will they understand technical terminology? A good basic approach for a technical report aimed at a wide range of professional audiences is **top-down structure**. Under top-down structure, readers get their first look at the experiment in the **title** and an **abstract**, which summarizes the most important experimental findings. Next, a more detailed description of the experiment, data summaries, and interpretations are presented in the body of the report in three separate sections: "**Materials and Methods**," "**Results**," and "**Discussion**." Finally, if the raw data (the ultimate details) is so voluminous as to detract from the body of the report, then it can be placed in an **appendix**. In your reports, we strongly recommend you use these sections as an overall structure. Sections should be clearly labeled with bold section headings, with blank space between sections. Details of the contents of each of these sections and some examples will be provided in this appendix.

TITLE AND ABSTRACT

The title and abstract are extremely important. You should word them very carefully. Some readers might be extremely busy people who have only a few minutes to look at your report. In a work setting, your immediate supervisor might have time to read it all, but his or her supervisor, who may be the president of the company, might not. In these times of information overload, a well-written title and abstract can quickly tell a reader the most essential outcomes of the experiment. He or she can then decide rationally whether to put the report down or read further to get more details. Get in the habit of questioning every word you use in the title and abstract. In the revision stage, play the role of a reader who has only five minutes: Will he or she get the "bottom line" results from only the title and abstract?

We suggest you center the title at the top of the first page, followed by the one-paragraph abstract at the middle of this page. At the bottom of this page put the date, author's name(s), and any other necessary information; for example, if some readers may not know the author, a mailing address may be appropriate. Though this page comes first in the report, it should be written last. The report itself has to exist before you can adequately summarize it.

Make the title as descriptive as possible, but keep it to one or two lines. For example, for Session 5 on real and perceived distances, the following might serve:

An Experiment Relating Guessed Distances and Actual Distances Between Visible Landmarks

Note the importance of the word *visible*, which efficiently tells the reader one important limitation of the experiment: that all the landmarks were near enough to be seen.

Without the word *visible*, the reader might think the subjects had been guessing the distance between the Statue of Liberty and the Eiffel Tower. Sometimes the title itself can be a "punchline" of the experimental results, as in the following:

Humans Tend to Underestimate Distances Between Visible Landmarks: Elementary Statistics Laboratory Session 5 Report

Good abstract writing is an art, unappreciated by most beginning writers. Figure A.1 contains two example abstracts for Session 5. Imagine the irritation of a supervisor reading the "not very good" abstract in Figure A.1, trying to get the gist of your experimental results in the minute he has before his next meeting. This abstract is not providing information; it is attempting to entice the reader into reading the rest of the report. This attempt will probably not succeed in the case of the manager, and the author will not be her favorite person for having provided the source of the irritation. Don't write an abstract designed purely to whet the appetite of the reader. The time for that is not in technical report writing.

The "better" abstract in Figure A.1 is much better. Notice the terse and factual tone and the specific numerical information. It is not possible to provide every detail of the experiment in one paragraph, but the "better" abstract gives the most important findings, very quickly and in relatively nontechnical language.

MATERIALS AND METHODS

This section is meant to tell the reader why the experiment was done, precisely how the data was collected, and what data analyses were done, without telling the results. Do not provide data summaries, graphs, tables, and so on in this section. Those belong in the "Results" section. The "Materials and Methods" section should begin with a clear statement of the purpose of the experiment. Next, a thorough and honest description of the manner in which data was collected should be presented. This description should include details of the use of measuring instruments, operational definitions of basic measurements, randomization schemes, blinding or double-blinding schemes, and so on.

This section should also contain a brief but complete description of data analysis methods, including the types of computer hardware and software used. Be careful to be complete in the description of statistical methods used; simply saying "a two-

Not Very Good

Abstract

Have you ever wondered whether you could guess distances accurately? We all have faith in our own judgment, yet there are reports that, in fact, human beings tend to underestimate distances between landmarks. An experiment was recently conducted to test this hypothesis. This report contains the surprising results of that experiment, so read on!

Better

Abstract

An experiment was conducted to relate 15 subjects' guessed distances to the true distances between 12 visible landmarks and a reference point. The true distances in the experiment varied from about 3 feet to about 300 feet. The results show that although humans tend to underestimate distances in this range by as much as 50 percent, there is surprising consistency within individuals. This means that an individual's guesses can often be corrected statistically to be quite accurate, usually within 10 percent of the true distance, using a calibration curve.

Figure A.1 Example Abstracts

sample t-test was performed" does not allow the reader to assess the validity of your methods. There are different kinds of two-sample t-tests one could perform, and what were the hypotheses (in words), anyway? Was it a one-sided or two-sided test? You might give references to textbooks that provide explicit formulas, but do not include formulas in the "Materials and Methods" section. If formulas are really necessary, they belong in an appendix.

Here is an example "Materials and Methods" section taken from a grade-A report written for "Secrets Behind a Green Thumb," the plant-growth experiment begun in Session 4 and concluded in Session 12 (this experiment and analysis may differ somewhat from your plant-growth experiment, if you do those sessions in your class).

MATERIALS AND METHODS

The purpose of this experiment was to investigate the effects of two different watering schemes on the growth rate of two varieties of peas. All computer work for this experiment was done using version 8.0 of Minitab for the Macintosh on a Macintosh Classic computer.

Each pea plant was grown in its own pot under one of four treatment conditions determined by combinations of two levels of each of two factors: variety (Early Maturing or America's Choice) and watering (1/3 cup or 1/2 cup, twice weekly). The four treatment combinations were:

A. Early Maturing with light watering

B. Early Maturing with moderate watering

C. America's Choice with light watering

D. America's Choice with moderate watering

Due to budget, time, and space constraints, only 3 pots were used for each of the 4 treatments. The 12 pots were arranged in random order on a windowsill in 6 aluminum tins, 2 per tin, each pot labeled according to its 2 factor levels. The random order in which the pots were placed on the sill was determined using Minitab's Integer Distribution function.

For the planting, each pot was filled with soil to 2 inches from the top, with a seed of the appropriate variety placed on the surface. Another inch of soil was then added, filling each pot to within 1 inch of the rim. Each pot was then watered with either 1/3 cup or 1/2 cup of ordinary tap water according to its designated watering treatment. During the course of the following eight weeks, a member of my team watered the plants on each Monday, and an instructor watered them on each Thursday. The plant's height was measured in centimeters each week by stretching it gently against a ruler.

After eight weeks had elapsed, the height data was summarized using Minitab. A worksheet was created with one variable (column) for each pot's height data. Row 1 corresponded to the heights after week 1, row 2 after week 2, and so on. A 13th variable, Time, was added, with values 1 through 8 corresponding to weeks 1 through 8. Four new variables were created by taking weekly averages of the heights of the three plants for each treatment combination. A "growth curve" for each treatment combination was then constructed, which was a scatter plot of mean heights versus week with the points connected. Then, four more scatter plots were created by overlaying these growth curves two at a time for easy comparison:

1. For only the Early Maturing variety, growth curves of mean heights under light and moderate watering

2. For only the America's Choice variety, growth curves of mean heights under light and moderate watering

3. For only the lightly watered plants, growth curves of mean heights for Early Maturing and America's Choice varieties

4. For only the moderately watered plants, growth curves of mean heights for Early Maturing and America's Choice varieties

Notice that this "Materials and Methods" section is very detailed about what work was done, but gives no indication of what results were obtained. The author gives details in turn on experimental design, data collection, and statistical methods. Notice the use of formatted lists to improve readability.

RESULTS

This section presents the experimental results in summaries, mostly tables and graphs, without interpretation. This allows readers to begin to form their own opinions and interpretations before reading yours, which come in the "Discussion" section. If anything unusual happened during data collection, it should be noted early in the "Results" section. The "Results" section should also contain a "road map" paragraph telling the reader where (in what tables and figures) the summary information is displayed. Tables should be numbered Table 1, Table 2, and so on, and graphs should be numbered Figure 1, Figure 2, and so on. Use these numbers to refer to tables and figures in the "Discussion" section.

It may seem a simple enough job to present a graph or a table, but often authors get sloppy or lazy and do not title and label tables and graphs adequately. This can make them as annoying to the reader as the "not very good" abstract in Figure A.1. The reader who has only one minute will read the abstract; the reader who has only three minutes will read the abstract and then look at the figures and tables. For this reason, a good rule of thumb is that, if possible, *a figure or table should stand alone as a piece of information*, without requiring reference to the text. For example, Figure A.2 shows two graphs from a Session 5 ("Real and Perceived Distances") trial run. The difference between the graphs is very obvious, and you may wonder why anyone would put an unlabeled graph in a report, but inexperienced authors do it all the time. The second figure has adequate labeling, including an informative title, units of measurement, and explanations of any lines or identified points on the graph. The second figure tells the reader a story in and of itself: that this class tends to underestimate distances (their guesses lie below the 45° line). Labeling for some graphs can be accomplished when they are constructed, such as with the Minitab Scatter Plot's Annotate subcommand, but most must be titled and labeled neatly by hand, with a typewriter, or by using the text tool of a drawing/painting software application.

Similar guidelines apply to the presentation of tables. They also should be informatively titled and labeled, including units of measurement. If the raw data can be presented in a table occupying less than half a page, it should be presented in the "Results" section. For nongraphical summaries of data, do not simply cut out computer output and paste it in the report. Usually computer output includes some extra information that can confuse readers and make it hard for them to find the most important summaries. Pick and choose the most appropriate summary measures (mean, standard deviation, etc.) from your computer output, and rearrange these measures neatly in a well-labeled table of your own design. Figure A.3 gives two examples of tables from Session 2, the heart-rate experiment.

The first table has simply been cut out of the Minitab Descriptive Statistics output and taped into the report. The title tells the reader nothing about what the data means. The variable names Rate15 and Rate30 are good for use in Minitab worksheets, but not for presentation to a reader unfamiliar with the experiment. The second table stands alone as a piece of information, as long as the reader knows what a mean, standard deviation, and range are. Note the subtitle giving identifying information about the subject, and the number of repetitions over which the summaries were calculated. Note also the Method column, which efficiently tells the reader how the measurements were made.

Not Very Good

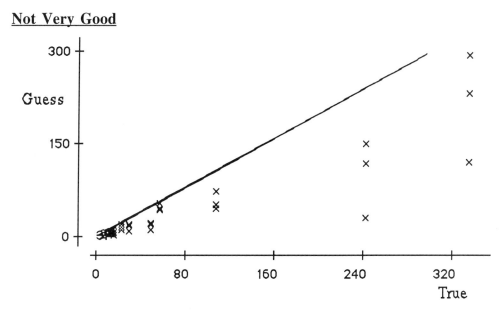

Better

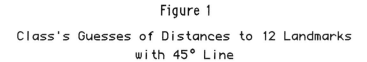

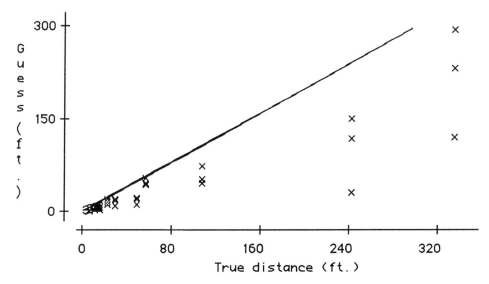

Figure A.2 Example Graphs

Not Very Good

Table 2. Heart-Rate Statistics

	N	MEAN	MEDIAN	TRMEAN	STDEV	SEMEAN
Rate15	8	79.75	76.25	77.50	7.74	2.83
Rate30	8	84.00	83.50	81.00	1.07	0.38

	MIN	MAX	Q1	Q3
Rate15	64	96	72.10	86.34
Rate30	80	88	82.05	85.77

Better

Table 2. Summaries of Heart Rate per Minute Measurements, Subject D.G.E. ($n = 8$ repetitions)

Method	Mean	Std. Dev.	Range
15-second count * 4	79.75	7.74	32
30-second count * 2	84.00	1.07	8

Figure A.3 Example Tables

DISCUSSION

In this final section of the report proper, honestly interpret the findings of the experiment and give any explanations you may have for patterns in the data. If the findings were inconclusive, or stand in contrast to what was expected in advance, say so. Point out any shortcomings or limitations to your interpretations. For example, in most of our sessions we are not able to truly obtain a random sample of human subjects, and this is a shortcoming. Also present any ideas and suggestions you may have for further experimentation. The scientific method is a cycle of hypothesis, experiment, hypothesis, experiment. Your experiment will usually generate new hypotheses, which you should state in the "Discussion" section.

In the presentation of formal statistical analyses (hypothesis tests and confidence intervals), do not speak "jargonese." For example, suppose the 95 percent confidence interval for the mean breaking strength in ounces of a brand of tissue was found to be (7.8, 8.2). A jargonese interpretation of this is, "We are 95 percent confident that μ lies within the interval (7.8, 8.2)." A clearer interpretation is, "We can be 95 percent confident that the true mean breaking strength of this brand of tissue, over a great many tests, would lie between 7.8 and 8.2 ounces." Nontechnical readers may prefer omission of the words "We can be 95 percent confident that." If this is done, an asterisk should follow the interpretation, and the footnote should then explain that the statement holds with 95 percent confidence and also what is meant by this.

APPENDIX

If the raw data is too voluminous to be included in a table of less than half a page, but not more than two or three pages, present it in a well-labeled table in an appendix. Also, if any equations or very technical material needs to be presented as clarification of the statistical methods used, these should be presented in an appendix.

APPENDIX TWO

Technical Report Writing Checklist

TITLE

Is the title one to two lines?

Is it fully descriptive?

Does every word count?

Are there better choices for any words?

ABSTRACT

Are all of the main experimental results presented?

Is the information specific rather than vague?

Does every word count?

Are there better choices for any words?

If this is all that the reader reads, will he or she know the bottom line?

MATERIALS AND METHODS

Are the goals of the experiment clearly described?

Are details of the sample selection process provided?

Are details of the basic measurement and the measuring equipment provided?

Are details of the experimental design presented?

Are the computer hardware and software described?

Is there a clear and correct description of the statistical methods?

Is there clear and correct use of terminology and notation?

Are the results or outcomes presented in the "Results" section and not here?

RESULTS

Are any unusual events that occurred during data collection noted?

Have you included a "road map" paragraph explaining the tables and figures?

If the raw data is not too voluminous, is it presented in a table?

Are all tables and figures numbered and well titled?

Are all tables and figures well labeled, including units of measurement?

Is there unnecessary or unclear information in tables or figures?

Will each table or figure be meaningful without reading the text?

Are all tables and figures neat and attractively presented?

DISCUSSION

Are the goals of the experiment addressed?

Are interpretations presented in understandable language (not "jargonese")?

Are any shortcomings or limitations of the experiment discussed?

Are ideas for further experimentation addressed?

APPENDIX (OPTIONAL)

Is the raw data presented in a well-titled and well-labeled table (if it is not in the report body)?

If equations are presented, is all notation clearly defined?

OVERALL

Have you proofread your report?

Is the spelling correct? Use a spelling checker, if possible.

Is the report free of grammatical errors?

Is the report unnecessarily wordy?

Have you checked for typographical errors in the numerical information?

Are the pages numbered?

Is the report clean and attractive?

INDEX

Alarm clock desk accessory, defined, 7
Alternative hypothesis, 248–250, 265
Analysis of variance, 181, 231

Bell shaped, 140, 146, 155–157, 159, 161
Bias, 80, 87–88, 93–94, 301–302, 304
Bimodal, 46, 146, 253
Binomial, 118, 123–124, 196
Biostatistician, 218
Blocking, 236, 238
Boxplot, 20, 30, 40–42, 109, 192, 256, 264–265, 269–270
Breaking strength, 50, 185–187, 189–190, 192–193, 197–200
By-treatment, 48
By-variable, 55–56, 139, 260, 263

Calculator desk accessory, defined, 7
Calibration, 80, 86–90, 94
Census, 256, 258
Central limit theorem, 148, 162–164, 166
Classification, 48–49, 58, 61–62, 65
Clicking, defined, 6
Clinical trial, 231
Close box, defined, 6
Concave, 86–87, 93–94
Confidence interval, 245, 247, 250–254, 259, 266–268, 270, 301, 315
  independent samples, 256
  one-sample t, 186, 193–196, 199–200, 246, 249, 251
  paired-sample t, 236
  proportion, 294, 301, 304
  sign, 186, 196–198, 200
  Wilcoxon, 197–198
Confidence statement, 245
Consumer price index, 112
Convenience sample, 258, 304
Convex, 86–87, 93–94
Correlation, 272, 278, 280, 282–283, 291
Cycles, 96–97, 99, 103, 108

Data window, defined, 11
Dependent sample, 251
Dependent variable, 69, 283
Desk accessories, 7
Desktop, defined, 5
Discriminant analysis, 49, 62

Diskette
  copying files between, 25, 42
  defined, 9
  destination, 25
  format, 10
  source, 25
Dotplot, 4, 20–22, 26–27, 40, 48, 55–57, 59, 65, 136, 173, 175, 192, 224, 244–245, 250, 253
Double-clicking, defined, 6
Dragging, defined, 7

Estimator, 148–149, 163
Experiment
  designed, 68, 75, 202, 212
  double-blind, 125
  factorial, 169
  one-factor, 181
  pair-t, 270
  planned, 68, 167–168, 181
  simulation, 163
  single-blind, 118–119, 125, 127
  two-factor, 68, 70, 169, 180
Experimental design, 231, 236, 311, 318
Experimental unit, 71, 202, 258
Exploratory data analysis, 109
Exponential, 150, 160–161, 166

Factor, 68–71, 75, 77–78, 93, 118, 125, 130, 168–169, 172–173, 175–178, 180–184, 202, 204, 212–213, 216, 218–219, 224–225, 227–228, 230–231, 233, 310
Federalist Papers, 62

Graph window, defined, 22
Growth curve, 207–208, 212–213, 310–311

Hard disk drive, defined, 5
Histogram, 20, 30, 40, 42, 45, 136, 153–155, 157, 159–163, 166, 192–193
History window, defined, 23
Hypothesis test, 194, 196, 249–250, 252, 254, 266–268, 270, 315
  Paired-sample t, 236

Icon, defined, 5
Independent samples, 238, 251, 265, 267
Independent variable, 69, 71, 283, 292
Info window, defined, 22

Interaction, 168, 176–178, 180, 184, 218, 225, 227–231, 233–234
Internal disk drive, 9

Keyboard command, defined, 19

Lagging, 105
Least squares, 283
Linear, 86, 94, 107, 212, 274, 280, 282–283, 286
Local control, 238

Main menu, defined, 5
Major axis, defined, 274
Mean, 16–21, 33–34, 37, 40–41, 44–45, 56–58, 107, 109–110, 115, 130, 136, 139, 140, 143, 146, 148–166, 168, 176–179, 181–184, 186, 193–194, 196–197, 199–200, 205–206, 215, 225, 227–230, 238, 246–248, 250–251, 253–254, 258–259, 263, 265, 267, 270, 277, 311–312, 315
Measurement error, 80, 141, 142, 145, 160, 162–163, 165–166
Median, 37, 41, 58, 83, 107–108, 110, 115, 148–166, 196–197, 199, 225
Minitab prompt, defined, 41
Minitab worksheet, defined, 12
Minor axis, defined, 274
Mode, 137
Mouse, defined, 4
Moving average, 103, 105–107, 110
Multiple regression, 288

Nonparametric, 186, 196, 200
Nonresponse bias, 302
Nonsymmetric, 149–150, 160, 163
Normal, 137, 150, 157–159, 162–163, 165–166, 186, 192–193, 196–200, 245, 301
Null hypothesis, 194–195, 200, 248–250, 265

Observed significance level, 249
Operational definition, 52, 63, 65, 132, 142, 145, 289, 298, 308
Outlier, 4, 21, 26–27, 46, 110, 130, 136, 137, 145, 253
Ozone hole, 26

P-value, 195, 249–250, 254
Paired sample, 236, 246, 251
Parameter, 148, 163

Percentile
   50th, 37
   75th, 37
   25th, 37
Point estimate, 301
Pointing, defined, 5
Polygon plot, 58–61, 65
Predictor variable, 283, 286, 288, 292
Preliminary study, 30, 43
Proportion, 118, 123, 125–127, 294, 298, 300–301
Protocol, 187, 189–190, 198

Quality 130, 142, 181, 186, 295
Quartile
   first, 37
   third, 37
Quitting Minitab, defined, 24

R-square, 283, 286
Randomization, 68, 70–71, 75, 78, 118, 168–169, 203, 213, 219, 231, 233, 308
Random sample, 122, 133, 257, 294–296, 300, 303, 314
Range, 38, 45–46, 312
Rate of occurrence, 54
Regression, 80, 87–88, 213, 272, 283–284, 286, 288, 292
Relative efficiency, 251–252, 254
Replicates, 223
Replication, 31, 70, 169
Resize box, defined, 8
Response variable, 69, 75, 77–78, 213, 283, 286, 288

Sample size, 30, 43, 161–163, 199, 300
Sampling, 129, 150, 293
   bias, 303
   distribution 130, 143, 146, 148–149, 151, 153–163, 165–166
   frame, 294, 296, 297, 301–302
Scatter plot, 48, 58-61, 80, 83–87, 89, 91, 93–94, 101–103, 108, 110, 138, 142, 145–146, 179, 202, 207–210, 215–216, 229, 272, 278–280, 282–283, 285, 291–292, 311
Scrolling in windows, defined, 14
Seasonal variation, 96–97, 108
Selecting menu items, defined, 7
Session window, defined, 21

Simple random sample, 133, 256, 258–259, 296
Simulation, 147–148, 150, 156, 158, 160, 163
Skewed, 46, 137, 146, 253
Smoothing, 97, 103, 107–109
Standard deviation, 33–34, 37, 45–46, 115, 130, 136, 142, 146, 155, 157, 159–160, 165, 199, 225, 300
Standard error, 300
Stem-and-leaf, 4, 20–22, 40, 109, 136–137, 140, 142, 145, 186, 191–193, 199
Symmetric, 46, 146, 149–150, 156, 158, 162–163, 165, 198, 253
Systematic random sample, 256, 258–259, 269

t, 150, 156–157, 162, 165, 245, 249
   Independent samples, 256, 267–268
   One-sample, 186, 194–196, 199, 246, 248
   Paired-sample, 236, 267
   Two-samples, 265–267, 269–270, 308–309
Taste test, 118, 125
Teamwork, 2
Threshold Test Ban Treaty, 89–90
Time series, 96–97, 101, 103–105, 112–113, 115

Title bar, defined, 8
Top-down structure, 306
Total quality management, 295
Transform, 288
Trash window, defined, 6
Treatment combination, 70–73, 169–170, 172–173, 175–176, 178, 181–182, 202–209, 212–213, 215–216, 219, 221–225, 228, 234, 310
Trends, 96–97, 99, 103, 108
Two-factor design, 68, 168, 202, 218

Unbiased, 87, 93
Uniform, 130, 150–152, 155–156, 158, 162, 165

Variability, 29–31, 38, 43–44, 49, 57–58, 69, 75, 80–81, 83–84, 103, 130–131, 136, 139, 141–143, 156, 158, 160, 162, 187, 190, 203, 209, 212, 215, 238, 265, 294, 300
Variance, 33–34, 251, 265–267, 270
Variation, 127–129, 132, 142, 162, 187, 271, 283, 286

Window, defined, 6
Working in teams, 1